Brainse Bhaile Thormod
Ballyfermot Library

Tel. 6269324/

WITHDRAWN FROM STOCK
DUBLIN CITY PUBLIC LIBRARIES

CONSCIOUS

CONSCIOUS

A Brief Guide to the Fundamental Mystery of the Mind

Annaka Harris

Brainse Bhaile Thormod
Ballyfermot Library

Tel. 6269324/

HARPER

An Imprint of HarperCollins*Publishers*

CONSCIOUS. Copyright © 2019 by Annaka Harris. All rights reserved. Printed in the United States of America. No part of this book may be used or reproduced in any manner whatsoever without written permission except in the case of brief quotations embodied in critical articles and reviews. For information, address HarperCollins Publishers, 195 Broadway, New York, NY 10007.

HarperCollins books may be purchased for educational, business, or sales promotional use. For information, please email the Special Markets Department at SPsales@harpercollins.com.

FIRST EDITION

Illustrations by Gemma O'Brien

Art direction by Jackie Phillips

Figures on pages 59, 70, 104, and 106 by ElectraGraphics, Inc.

Library of Congress Cataloging-in-Publication Data has been applied for.

ISBN 978-0-06-290671-7

19 20 21 22 23 LSC 10 9 8 7 6 5 4 3

For Sam, Emma, and Violet

Contents

CONSCIOUS

1

A MYSTERY HIDING IN PLAIN SIGHT

Our experience of consciousness is so intrinsic to who we are, we rarely notice that something mysterious is going on. Consciousness is *experience itself*, and it is therefore easy to miss the profound question staring us in the face in each moment: Why would any collection of matter in the universe be conscious? We look right past the mystery as if the existence of consciousness were obvious or an inevitable result of complex life, but when we look more closely, we find that it is one of the strangest aspects of reality.

Thinking about consciousness can spark the same kind of pleasure we get from contemplating the nature of time or the origin of matter, invoking a deep curiosity about ourselves and the world around us. I remember looking up at the sky when I was young and realizing that my usual sense of being down on the ground with the sky above me wasn't an entirely accurate perception. I was intrigued by the fact that even though I had learned that gravity pulls us toward the earth as we orbit the sun—and that there is no real "up" and "down"—my *feeling* of being down on the ground below the sky had remained unchanged. To shift my perspective, I would sometimes lie outside with my arms and legs outstretched and take in as much of the sky and horizon as possible. Attempting to break free of the familiar feeling of being *down here* with the moon and stars above me, I would relax all my muscles—surrendering to the force holding me tightly to the surface of our planet—and focus on the truth of my situation: *I'm floating around the universe on this giant sphere—suspended here by gravity and going for a ride.* Lying there, I could sense that I was in fact looking *out* at the sky, rather than *up*. The delight I experienced came from temporarily silencing a false intuition and glimpsing a deeper truth: being on the earth doesn't separate us from the rest of the universe; indeed, we are and have always been in outer space.

This book is devoted to shaking up our everyday assumptions about the world we live in. Some facts are so important and so counterintuitive (matter is mostly made up of empty space; the earth is a spinning sphere in one of billions of solar systems in our galaxy; microscopic organisms cause disease; and so on) that we need to recall them again and again, until they finally permeate our culture and become the foundation for new thinking. The fundamental mysteriousness of consciousness, a subject deeply perplexing to philosophers and scientists alike, holds a special place among such facts. My goal in writing this book is to pass along the exhilaration that comes from discovering just how surprising consciousness is.

Before posing any questions about consciousness, we

must determine what we are talking about in the first place. People use the word in a variety of ways; for example, in referring to a state of wakefulness, a sense of selfhood, or the capacity for self-reflection. But when we want to single out the mysterious quality at the heart of consciousness, it's important to narrow in on what makes it unique. The most basic definition of consciousness is that given by the philosopher Thomas Nagel in his famous essay "What Is It Like to Be a Bat?," and it is how I use the word throughout this book. The essence of Nagel's explanation runs as follows:

An organism is conscious if there is *something that it is like* to be that organism.[1]

In other words, consciousness is what we're referring to when we talk about experience in its most basic form. Is it *like something* to be you in this moment? Presumably your answer is yes. Is it *like something* to be the chair you're sitting on? Your answer will (most likely) be an equally definitive no. It's this simple difference—whether there is an experience present or not—which we can all use as a reference point, that constitutes what I mean by the word "consciousness." Is it *like something* to be a grain of sand, a bacterium, an oak tree, a worm, an ant, a mouse, a dog? At some point along the spectrum the answer is yes, and the great mystery lies in why the "lights turn on" for some collections of matter in the universe.

We can even wonder: At what point in the development of a human being does consciousness flicker into existence? Imagine a human blastocyst just a few days old, consisting of only about two hundred cells. We assume there is probably *nothing it is like* to be this microscopic collection of cells. But over time these cells multiply and slowly become a human baby with a human brain, able to detect changes in light and recognize its mother's voice, even while in the womb. And, unlike a computer, which can also detect light and recognize voices, this processing is accompanied by an *experience* of light and sound. At whatever point in the development of a baby's brain your intuition tells you, *OK, now an experience is being had in there*, the mystery lies in the transition. First, as far as consciousness is concerned, there is nothing, and then suddenly, magically, at just the right moment . . . *something.* However minimal that initial something is, experience apparently ignites in the inanimate world, materializing out of the darkness.

After all, an infant is composed of particles indistinguishable from those swirling around in the sun. The particles that compose your body were once the ingredients of countless stars in our universe's past. They traveled for billions of years to land here—in this particular configuration that is you—and are now reading this book. Imagine following the life of these particles from their first appearance in space-time to the very moment they became arranged in such a way as to start *experiencing* something.

The philosopher Rebecca Goldstein paints a wonderfully clear and playful portrait of the mystery:

> Sure, consciousness is a matter of matter—what else could it be, since that's what we *are*—but still, the fact that some hunks of matter have an inner life . . . is unlike any other properties of matter we have yet encountered, much less accounted for. The laws of matter in motion can produce *this*, all *this*? Suddenly, matter wakes up and takes in the world?[2]

The moment matter becomes conscious seems at least as mysterious as the moment matter and energy sprang into existence in the first place. The mystery of consciousness rivals one of the greatest conundrums ever to bend human thought: How could something appear out of nothing?[3] Likewise, how does felt experience arise out of nonsentient matter? The Australian philosopher David Chalmers famously termed this the "hard problem" of consciousness.[4] Unlike the "easy problems" of explaining animal behavior or understanding which processes in the brain give rise to which functions, the hard problem lies in understanding why some of these physical processes have an experience associated with them at all.

Why do certain configurations of matter cause that matter to light up with awareness?

2

INTUITIONS AND ILLUSIONS

Now that we have a working definition of consciousness and the mystery it entails, we can start chipping away at some common intuitions. In large part, our intuitions have been shaped by natural selection to quickly provide life-saving information, and these evolved intuitions can still serve us in modern life. For example, we have the ability to unconsciously perceive elements in our environment in threatening situations that in turn deliver an almost instantaneous assessment of danger—such as an intuition that we shouldn't get into an elevator with someone, even though we can't put our finger on why. Your brain is often processing helpful cues you may not be aware of in the moment: the other person who is getting into the elevator is flushed or has dilated pupils (both are signals that he is adrenalized and about to act violently), or the door to the building, which is usually locked, has been left ajar. We can know that a situation is dangerous without having any idea how or why we know it. Our intuitions are also shaped through learning, culture, and other environmental factors. We sometimes have useful intuitions about life decisions—such as which apartment to rent—born of relevant information that our brain has acquired, and taken into account, through unconscious processes. In fact, research suggests that our "gut feelings" are more reliable in many situations than the fruits of conscious reasoning.[1]

But our gut can deceive us as well, and "false intuitions" can arise in any number of ways, especially in domains of understanding—such as science and philosophy—that evolution could never have foreseen. Consider probability and statistics, where our intuitions are notoriously unreliable: Many of us are nervous fliers, despite the fact that, statistically, we would need to fly every day for about 55,000 years before being involved in a fatal plane crash (and it's worth mentioning that although people don't commonly have panic attacks when getting behind the wheel in preparation for a trip to the grocery store, one's safety on such trips is actually less secure by many orders of magnitude than while flying).[2] We can barely square our intuitions with some of the most basic scientific facts—the earth seemed flat to us until breakthroughs in celestial observations revealed otherwise. And in some areas of study, such as quantum physics, our intuitions are not only useless but are an outright obstacle to progress. An intuition is simply the powerful sense that something is true without our having an awareness or an understanding of the reasons behind this feeling—it may or may not represent something true about the world.

In this chapter, we will consider our intuitions regarding

how we judge whether or not something is conscious, and we'll discover that the seemingly obvious answers sometimes fall apart on closer inspection. I like to begin this exploration with two questions that at first glance appear deceptively simple to answer. Note the responses that first occur to you, and keep them in mind as we explore some typical intuitions and illusions.

1. In a system that we know has conscious experiences—the human brain—what evidence of consciousness can we detect from the *outside*?
2. Is consciousness essential to our behavior?

These two questions overlap in important ways, but it's informative to address them separately. Consider first that it's possible for conscious experience to exist without any outward expression at all (at least in a brain). A striking example of this is the neurological condition called locked-in syndrome, in which virtually one's entire body is paralyzed but consciousness is fully intact. This condition was made famous by Jean-Dominique Bauby, the late editor in chief of French *Elle*, who ingeniously devised a way to write about his personal story of being "locked in." After a stroke left him paralyzed, Bauby retained only the ability to blink his left eye. Amazingly, his caretakers noticed his efforts to communicate, and over time they developed a method whereby he could spell out words through a pattern of blinks, thus revealing the full scope of his conscious life. He describes this harrowing experience in his 1997 memoir, *The Diving Bell and the Butterfly*, which he wrote in about two hundred thousand blinks. Of course, we may assume that his consciousness would not have been changed whatsoever if his left eyelid had succumbed to the paralysis as well. And without this mobility, there would have been absolutely no way for him to communicate that he was fully conscious.

Another example of bodily imprisonment is a condition called "anesthesia awareness," in which a patient given a general anesthetic for a surgical procedure experiences only the paralysis without losing consciousness. People in this condition must live out the nightmare of feeling every aspect of a medical procedure, sometimes as drastic as the removal of an organ, without the ability to move or communicate that they are fully awake and experiencing pain. This and the previous example seem to come straight out of a horror movie, but we can imagine other, less disturbing instances in which a conscious mind might lack a mode of expression—scenarios involving artificial intelligence (AI), for example, in which an advanced system becomes conscious but has no way of convincingly communicating this to us. But one thing is certain: it's possible for a vivid experience of consciousness to exist undetected from the outside.

Now let's go back to the first question and ask ourselves: What might qualify as evidence of consciousness? For the most part, we believe we can determine whether or not an organism is conscious by examining its behavior. Here is a simple assumption most of us make, in line with our intuitions, that we can use as a starting point: "People are conscious; plants are not conscious." Most of us feel strongly that this statement is correct, and there are good scientific reasons for believing that it is. We assume that consciousness does not exist in the absence of a brain or a central nervous system. But what evidence or behavior can we observe to support this claim about the relative experience of human beings and plants? Consider the types of behavior we usually attribute to conscious life, such as reacting to physical harm or caring for others. Research reveals that plants do both of these things in complex ways—though, of course, we conclude that they do so without feeling pain or love (i.e., without consciousness). But some behaviors of people and plants are so alike that this in fact poses a challenge to our using certain behavior as evidence of conscious experience.

In his book, *What a Plant Knows: A Field Guide to the Senses*, Daniel Chamovitz describes in fascinating detail how the stimulation of a plant (by touch, light, heat, etc.) can cause reactions similar to those in animals under analogous conditions. Plants can sense their environment through touch and can detect many aspects of their surroundings, including temperature, by other modes. It's actually quite common for plants to react to touch: a vine will increase its rate and change its direction of growth when it senses an object nearby to wrap itself around, and the infamous

Venus flytrap can distinguish between a heavy rain or a strong gust of wind, which do not cause its blades to close, and the tentative incursions of a nutritious beetle or frog, which will make them snap shut in one-tenth of a second.

Chamovitz explains how the stimulation of a plant cell causes cellular changes that result in an electrical signal—similar to the reaction caused by the stimulation of nerve cells in animals—and "just like in animals, this signal can propagate from cell to cell, and it involves the coordinated function of ion channels including potassium, calcium, calmodulin, and other plant components."[3] He also describes some of the shared mechanisms between plants and animals down to the level of DNA. In his research, Chamovitz discovered which genes are responsible for a plant's ability to determine whether it's in the dark or the light, and these genes, it turns out, are also part of human DNA. In animals, these same genes regulate responses to light and are involved in "the timing of cell division, the axonal growth of neurons, and the proper functioning of the immune system." Analogous mechanisms exist in plants for detecting sounds, scents, and location, and even for forming memories. In an interview with *Scientific American*, Chamovitz describes how different types of memory play a role in plant behavior:

> If memory entails forming the memory (encoding information), retaining the memory (storing information), and recalling the memory (retrieving in-

> formation), then plants definitely remember. For example a Venus Fly Trap needs to have two of the hairs on its leaves touched by a bug in order to shut, so it remembers that the first one has been touched. . . . Wheat seedlings remember that they've gone through winter before they start to flower and make seeds. And some stressed plants give rise to progeny that are more resistant to the same stress, a type of transgenerational memory that's also been recently shown also in animals.[4]

The ecologist Suzanne Simard conducts research in forest ecology, and her work has produced breakthroughs in our understanding of intertree communication. In a 2016 TED talk, she described the thrill of uncovering the interdependence of two tree species in her research on mycorrhizal networks—elaborate underground networks of fungi that connect individual plants and transfer water, carbon, nitrogen, and other nutrients and minerals. Simard was studying the levels of carbon in two species of tree, the Douglas fir and the paper birch, when she discovered that the two species were engaged "in a lively two-way conversation." In the summer months, when the fir needed more carbon, the birch sent more carbon to the fir; at other times when the fir was still growing but the birch needed more carbon because it was leafless, the fir sent more carbon to the birch—revealing that the two species were in fact interdependent. Equally surprising were the results of further research led by Simard, showing that Douglas fir "mother trees" were able to distinguish between their own kin and a neighboring stranger's seedlings. Simard found that the mother trees colonized their kin with bigger mycorrhizal networks, sending them more carbon belowground. The mother trees also "reduced their own root competition to make room for their kids," and, when injured or dying, sent messages through carbon and communicated other defense signals to their kin seedlings, increasing the seedlings' resistance to local environmental stresses.[5] Likewise, by spreading toxins through underground fungal networks, plants are also able to fight off threatening species. Because of the vast interconnections and functions of these mycorrhizal networks, they have been referred to as "earth's natural Internet."[6]

Still, we can easily imagine plants exhibiting the behaviors described here without there being *something it is like* to be a plant, so complex behavior doesn't necessarily shed light on whether a system is conscious or not. We can probe our intuitions about behavior from another angle by asking, "Does a system need consciousness to exhibit certain behaviors?" For instance, would an advanced robot need to be conscious to give its owner a pat on the back when it witnessed her crying? Most of us would probably answer, "Not necessarily." At least one tech company is creating computerized voices indistinguishable from human ones.[7] If we design an AI that one day begins saying things like, "Please stop—it hurts when you do that!" should we take this as evidence of consciousness, or simply of complex programming in which the lights are off?

We assume, for example, that an entirely nonconscious algorithm is behind Google's growing ability to accurately guess what we are searching for, or behind Microsoft Outlook's ability to make suggestions about whom we might want to cc on our next email. We don't think our computer is conscious, much less that it cares about us, when it flashes Uncle John's contact, reminding us to include him in the baby announcement. The software has obviously learned that Uncle John usually gets included in emails to Dad and Cousin Jenny, but we never have the impulse to say, "Hey, thanks—how thoughtful of you!" It's conceivable, however, that future deep-learning techniques will enable these machines to express seemingly conscious thoughts and emotions (giving them increased powers to manipulate people). The problem is that both conscious and nonconscious states seem to be compatible with any behavior, even those associated with emotion, so a behavior itself doesn't necessarily signal the presence of consciousness.

Suddenly, our reflexive answers to question 1—What constitutes evidence for consciousness?—are beginning to dissolve. And this leads us to question 2, regarding whether consciousness performs an essential function in—or has any effect at all on—the physical system that's conscious.[8] In theory, I could act in all the ways I do and say all the things I say without having a conscious experience of it, much as an advanced robot might (though, admittedly, it's hard to imagine). This is the gist of a thought experiment referred to as the "philosophical zombie," which was made popular by David Chalmers. Chalmers asks us to imagine that any person could, in effect, be a zombie—someone who looks and acts exactly like everyone else on the outside without experiencing anything at all on the inside. The zombie thought experiment is controversial, and other philosophers, notably Daniel Dennett of Tufts University, claim that what it proposes is impossible—that a fully functioning human brain must be conscious, by definition. But the conceivability of a "zombie" is worth contemplating if only in theory, because it helps us pin down which behaviors, if any, we think *must* be accompanied by consciousness.

The goal here is to pry loose as many false assumptions as possible, and this particular mental exercise is useful whether or not a zombie is compatible with the laws of nature. Imagine that someone in your life is in fact an unconscious zombie or an AI (it could be anyone from a stranger behind a store counter to a close friend). The moment you witness behavior in this person that you think must coincide with an internal experience, ask yourself why. What role does consciousness seem to play in his behavior? Let's say your zombie friend witnesses a car accident, looks appropriately concerned, and takes out his phone to call for an ambulance. Could he possibly be going through these motions without an experience of anxiety and concern, or without a conscious thought process that leads him to make the call and describe what happened? Could this all take place even if he were a robot, without a felt experience prompting the behavior at all?

I have discovered that the zombie thought experiment is also capable of influencing our thinking beyond its intended function. Once we imagine human behavior around us existing without consciousness, that behavior begins to look more like many behaviors we see in the natural world that we've always assumed were nonconscious, such as the obstacle-avoiding behavior of a starfish, which has no central nervous system.[9] In other words, when we trick ourselves into imagining that people lack consciousness, we can begin to wonder if we're in fact tricking ourselves *all the time* when we deem other living systems—climbing ivy, say, or stinging sea anemones—to be without it. We have a deeply ingrained intuition, and therefore a strongly held belief, that systems that act like us are conscious, and those that don't are not. But what the zombie thought experiment makes vivid to me is that the conclusion we draw from this intuition has no real foundation. Like a 3-D image, it collapses the moment we take our glasses off.

3

IS CONSCIOUSNESS FREE?

As we go about our daily lives, we experience what appears to be a continuous stream of present-moment events, yet we actually become conscious of physical events in the world slightly *after* they have occurred. In fact, one of the most startling findings in neuroscience has been that consciousness is often "the last to know." Visual, auditory, and other kinds of sensory information move through the world (and our nervous system) at different rates. The light waves and sound waves emitted the moment the tennis ball makes contact with your racket, for example, do not arrive at your eyes and ears at the same time, and the impact felt by your hand holding the racket occurs at yet another interval. To complicate matters further, the signals perceived by your hands, eyes, and ears travel different distances through your nervous system to reach your brain (your hands are a lot farther away from your brain than your ears are). Only after all the relevant input has been received by the brain do the signals get synchronized and enter your conscious experience through a process called "binding"—whereby you see, hear, and feel the ball hit the racket all in the same instant. As the neuroscientist David Eagleman puts it:

> Your perception of reality is the end result of fancy editing tricks: the brain hides the difference in arrival times. How? What it serves up as reality is actually a delayed version. Your brain collects up all the information from the senses before it decides upon a story of what happens. . . . The strange consequence of all this is that you live in the past. By the time you think the moment occurs, it's already long gone. To synchronize the incoming information from the senses, the cost is that our conscious awareness lags behind the physical world.[1]

Surprisingly, our consciousness also doesn't appear to be involved in much of our own behavior, apart from bearing witness to it. A number of fascinating experiments have been conducted in this area, and the neuroscientist Michael Gazzaniga describes some of them in detail in a wonderful chapter aptly titled "The Brain Knows Before You Do" in his book *The Mind's Past*. Some of these experiments, most famously conducted by Benjamin Libet at the University of California, San Francisco, show that your brain prepares a complex motor movement of your body before you are consciously aware of deciding to move. In such experiments, subjects watch a special clock and, according to an instrument similar to the second hand on a traditional clock, mark the exact moment they decide to move (a finger, for instance). But, using EEG, researchers can reliably detect the cortical activity signaling these impending movements about half a second *before subjects feel they make the decision to move.*[2] More sophisticated versions of these experiments have been conducted since, producing the same results.[3] Though it's not clear how these types of simple motor decisions relate to more complex decisions, like choosing what to eat for lunch or deciding between two job offers, there is no question that modern neuroscience is providing us with a quickly evolving view of the human mind. We now have reason to believe that with access to certain activity inside your brain, another person can know what you're going to do before you do.

Our intuition that consciousness is behind certain behaviors is informed by our experience of freely making choices in the world, as our willed actions are inextricably linked to a sense of conscious control in the present moment. Whether it's as minor a decision as choosing water over orange juice or as consequential as taking the job in Texas instead of the one in New York, we feel strongly that consciousness is required for the thought processes (and even the preferences) involved in making a decision. Consequently, findings about how decisions are made at the level of the brain—and the milliseconds of delay in our conscious awareness of sensory input and even of our own thoughts—have caused many neuroscientists, Gazzaniga included, to describe the feeling of conscious will as an illusion. Note that in such experiments, the subjects felt they were making a freely willed action that, in actuality, had already been set in motion before they felt they made the decision to move.

The argument that conscious will is an illusion is further strengthened by the fact that this illusion can be intentionally triggered and manipulated. Experimenters have been able to cause a feeling of will in subjects when the subjects in fact had no control. It seems that, under the right conditions, it's possible to convince people that they have consciously initiated an action that was actually controlled by someone else. A series of such studies were conducted by the psychologists Daniel Wegner and Thalia Wheatley. Wegner explains:

> We have a participant in the experiment put their hands on a little board that's resting on top of a computer mouse, and the mouse moves a cursor around on a screen. The screen has a variety of different objects, pictures from the book *I-Spy*—in this case little plastic toys. Also in the room is our confederate; both of them have headphones on, and together they are asked to move the cursor around the screen and rest on an object every few seconds, whenever music comes on. . . . Most of the time they hear sounds over the headphones they're wearing, and some of these are names of things on the screen. The key part of the experiment occurs when, in some trials, the

> confederate is asked to force our subject to land the cursor on a particular object, so the person who we're testing hasn't done it, but has been forced. It's just as though someone was cheating on a Ouija board. We play the name of the object to our participant at some interval of time before or after they're forced to move, and we find that if we play the name of the object just a second before they're forced to move to it, they report having done it intentionally. . . . The feeling of agency can be fooled—and yet, we go about our daily lives feeling the opposite.[4]

So what role does consciousness play if it's not creating the will to move but merely watching the movement play out, all the while under the illusion that it is involved? We can see how the feeling of free will, as we typically experience it, is not as straightforward as it seems. And if we dispel this common notion, we can begin to question the idea that consciousness plays an integral role in guiding human behavior.

It's important to clarify that when I talk about the nature of free will in this context, I'm referring specifically to the feeling of a *conscious* will. I'm pointing to the fundamental, day-to-day illusion we all seem to walk around with: that we are distinct and separate "selves," separate not only from those around us and from the outside world but even from our own bodies, as if our conscious experience somehow floats free of the material world. For example, like everyone else, I have the absurd tendency to regard "my body" (including "my head" and "my brain") as something my conscious will inhabits—when in fact everything I think of as "me" is dependent on the functioning of my brain. Even the slightest neural changes, via intoxication, disease, or injury, could render "me" unrecognizable. Yet I can't seem to shake the false intuition that I could even choose to leave my body (if I could only figure out how) and everything constituting "me" would somehow remain magically intact. It's easy to see how human beings across the globe, generation after generation, have effortlessly constructed various notions of a "soul" and descriptions of life after death that bear a striking resemblance to life before death.

The brain, as a system, does have a type of free will, however—in that it makes decisions and choices on the basis of outside information, internal goals, and complex reasoning. But when I discuss the illusion of conscious will here, I'm speaking of the illusion that *consciousness is the will itself.*[5] The concept of a conscious will that is free seems to be incoherent—it suggests that one's will is separate and isolated from the rest of its environment, yet paradoxically able to influence its environment by making choices within it.

I was once at an event where my friend and meditation teacher Joseph Goldstein was asked if he believed we have free will. He answered the question with arresting clarity when he said that he couldn't even figure out what the term could possibly mean. What does it mean to have a will that is free from the cause-and-effect relationships of the universe? As he gestured with his hands dancing above him in the air, trying to point to this imaginary free will, he asked, "How can we even try to picture such a will floating about?"

Many people, however, object on ethical grounds to the assertion that conscious will is an illusion, holding that people should be held responsible for their choices and behavior. But people can (and should) be held responsible for their actions, for a variety of reasons; the two beliefs are not necessarily contradictory. We can still acknowledge the difference between premeditated, lucid actions and the sort that are caused by mental illness or other disorders of the mind/brain.[6]

Imagine we're in a future city, and a self-driving car hits a pedestrian. The response to this unfortunate event would depend on why the car didn't stop. If it turns out its software is flawed and can't detect pedestrians when they are bundled up in dark winter coats, for instance, that would require one response. If the car's sensors malfunctioned due to a defect specific to that one particular car, that would require a different response. And if the car hit the pedestrian because it was avoiding colliding with a crowded bus and pushing it into oncoming traffic, we would view this situation (and respond to it) very differently from the first two scenarios—as a "success" of the car's advanced technology, rather than a flaw. Simply knowing that a self-driving car hit a pedestrian isn't enough information to help us stop this car from becoming a repeat offender or to learn how to build better cars.

It's important to notice that in these reflections about self-driving cars, consciousness never entered the conversation. And the brain can be viewed in an analogous way when it comes to conscious will. Knowing *why* someone has behaved violently, for instance, will always be relevant. There are a range of human behaviors that can be influenced by deterrence, negative consequences, and empathy, along with inculcating the developing brains of children with self-regulation and self-control—and all the other methods civilized societies use to keep human beings (generally) well behaved.

The brain continually alters its behavior in response to input. It also changes and develops through memory, learning, and internal reasoning. With the proper guidance, we eventually stop collapsing onto the floor and pounding our fists when we don't get our way. We couldn't accomplish this without concepts such as responsibility, accountability, and consequences. But in situations in which the usual civilizing pressures are powerless (when someone is suffering from schizophrenic hallucinations, for instance), it makes sense to treat that person and his behavior differently from someone subject to those pressures. Similarly, understanding the intentions behind violent behavior gives us relevant information about what kind of "software" someone's brain is running. A person who plots multiple murders has a brain that is operating very differently from someone who has a stroke while driving and accidentally kills a number of people.

It may seem paradoxical to talk about ethics in this framework, because consciousness is essential to ethical questions. As a domain that pertains to suffering, all conversations about ethics are about how something *feels*. But in terms of the brain being a physical processing system, some of its goals can be ethical in nature—namely, minimizing the number of events that cause suffering—and here our brains are analogous to the self-driving cars mentioned earlier. Even though we are talking about modifying a conscious experience, consciousness itself isn't necessarily controlling the system; all we know is that consciousness is experiencing the system. It is no contradiction to say that consciousness is essential to ethical concerns, yet irrelevant when it comes to will.

A distinction between the brain's intentional behaviors and behaviors that are caused by brain damage or other outside forces ("against one's will") is valid and necessary, especially when structuring a society's laws and criminal justice systems. But the claim that conscious will is illusory still stands—in the sense that consciousness is not steering the ship—and can be maintained alongside these other distinctions of intentionality and responsibility.

The experiments described in this chapter are in fact not necessary to prove the point. Our experience alone reveals the illusion, and you can gain insight into this with a simple experiment. Sit in a quiet place and give yourself a choice—to lift either your arm or your foot—that must be made before a given time (before the second hand on the clock reaches the six, for example). Do this over and over again and observe your moment-to-moment experience closely. Notice how this choice gets made in real time and what it feels like. Where does the decision come from? Do you *decide when to decide*, or does a decision simply arise in your conscious experience? Does a free-floating conscious will somehow deliver the thought, *Move your arm*, or is the thought delivered to you? What actually made you choose arm over foot? It may suddenly seem that "you" (meaning your conscious experience) didn't have any part in it.

It seems clear that we can't decide what to think or feel, any more than we can decide what to see or hear. A highly complicated convergence of factors and past events—including our genes, our personal life history, our immediate environment, and the state of our brain—is responsible for each next thought. Did you decide to remember your high school band when that song started playing on the radio? Did I decide to write this book? In some sense, the answer is yes, but the "I" in question is not my conscious experience. In actuality, my brain, in conjunction with its history and the outside world, decided. I (my consciousness) simply witness decisions unfolding.

4

ALONG FOR THE RIDE

Another wellspring of examples that turn our intuitions upside down and challenge the typical notion of free will can be found in the study of parasites and how they affect the behavior of their hosts. *Toxoplasma gondii* is a microscopic parasite that can infect all warm-blooded animals but can sexually reproduce only in the intestines of a cat. While it can survive in any mammal, it must eventually make its way back to a feline to complete its life cycle. *Toxoplasma* most commonly infects rats, because they frequent many of the same hangouts as cats, and the parasite has evolved a brilliant and extremely creepy mechanism for overcoming the challenge of traveling from the rats, who have a deeply ingrained fear of cats, back to its reproductive home. By a neurological mechanism that scientists still don't completely understand, *Toxoplasma* affects the behavior of the infected rats, causing them to forsake their fear of cats and in many cases to walk (or even run) directly toward their enemy. *Toxoplasma* creates hundreds of cysts in the brain of its host, causing dopamine levels to rise. Dopamine is a neurotransmitter that plays a role in mediating powerful emotions such as desire and fear, which helps explain much of the behavior we see in mammals infected with the parasite. It's possible that these rats somehow feel that they are being manipulated against their will by an outside force, but it seems more likely that their neurochemistry is being altered and thus their desires and fears change: they no longer feel afraid of cats and are now, in fact, drawn to them.[1]

Humans can become infected with the parasite in the same way other mammals can—by consuming the undercooked meat of infected animals or by coming in direct contact with environments contaminated with cat feces, such as drinking water, garden soil, or litter boxes—and it turns out that *Toxoplasma* also has an effect on human brains. As the science journalist Kathleen McAuliffe reports on observations made by parasitologists, "Neurons harboring the parasite were making 3.5 times more dopamine. The chemical could actually be seen pooling inside infected brain cells." *Toxoplasma* can cause a variety of behavioral changes in humans and is thought to be a trigger for schizophrenia and other mental illnesses in many people. According to McAuliffe's reporting, "people with schizophrenia are two to three times more likely to test positive for antibodies to the parasite than those who don't have the disorder."[2]

In her fascinating and amusing *New York Times* article "In Parasite Survival, Ploys to Get Help from a Host," Natalie Angier reports:

> When Jaroslav Flegr of Charles University in Prague administered personality tests to two groups of people, one showing immunological signs of a prior Toxoplasma infection and the other not, infected men scored comparatively higher than uninfected men in traits like suspicion of authority and a propensity to break rules, while infected women ranked relatively higher than noninfected women in measures of warmth, self-assurance and chattiness.

There are countless examples of other parasites affecting the behavior of their hosts. A horsehair worm will cause an infected cricket, which would normally maintain a safe distance from large bodies of water, to race toward the nearest lake or stream. By releasing neurochemicals that mimic those of a cricket, the worm urges the cricket to plunge in just in time for the worm to participate in mating season, which must take place in the water.[3] Similarly, although pill bugs usually hide out during daylight hours to avoid being eaten by birds, those that are infected with the thorny-headed worm want nothing more than to venture out for a nice afternoon of sunbathing—on a light-colored surface, no less, where the high-contrast environment makes them easy to spot by birds flying overhead. The worms then hitch a ride back to the bird's digestive system to lay their eggs.[4] The larvae of the Alcon blue butterfly have a surface chemistry that mimics chemicals found on the surface of at least two species of ant larvae, causing the ants to carry the familiar-scented butterfly larvae back to their nest to feed and nurture them, often at the expense of their own offspring.[5] And parasitic wasps cause orb spiders to build webs that differ drastically from their usual design. After the wasp larva injects a chemical into the spider, the spider begins spinning a web much more suited to the larva's needs than its own, keeping the larva safe from nearby predators and providing the perfect netting for building its cocoon.[6] The list goes on and on.

When reviewing examples like these, we are immediately struck by how often we are blind to the complex array of forces at play in the behavior taking place all around us. One can't help but wonder what's truly driving all our own desires and personality traits—especially ones we tend to strongly identify with.

There are also instances of bacterial infections causing behavioral changes in people, and scientists are continuing to discover links between infections and human psychological disorders.[7] *Streptococci* bacteria, for instance, have evolved a defense mechanism enabling them to hide successfully from the immune system of children for some period. Molecules on the walls of their cells make them indistinguishable from tissues in a child's heart, joints, skin, and brain. Eventually the child's immune system recognizes the strep as foreign to the body, but when it launches its attack, it may mistakenly target healthy tissues in the body as well. According to studies at the National

Institute of Mental Health, in these cases "some cross-reactive 'anti-brain' antibodies [may] target the brain, causing OCD, tics, and the other neuropsychiatric symptoms of PANDAS [Pediatric Autoimmune Neuropsychiatric Disorders Associated with Streptococcal infections]."[8] Here, the behavior of the host isn't supporting the goals of the parasite; rather, the strep infection results in a phenomenon with "unintended" effects. But both types of examples uncover the same reality about our conscious experience, and the idea that "I" am the ultimate source of my desires and actions begins to crumble.

With so many behind-the-scenes forces at work—from the essential neurological processes we previously examined to bacterial infections and parasites—it's hard to see how our behavior, preferences, and even choices could be under the control of our conscious will in any real sense. It seems much more accurate to say that consciousness is along for the ride—watching the show, rather than creating or controlling it. In theory, we can go as far as to say that few (if any) of our behaviors need consciousness in order to be carried out. But at an intuitive level, we assume that because human beings act in certain ways *and* are conscious—and because experiences such as fear, love, and pain feel like such powerful motivators within consciousness—our behaviors are driven by our *awareness of them* and otherwise would not occur. However, it's now obvious that many behaviors we usually attribute to consciousness, and think of as proof of consciousness, could actually exist without consciousness, at least in theory. This brings us back to our two questions. And, once again, it's hard to see how conscious experience plays a role in behavior. That's not to say it doesn't, but it's almost impossible to point to specific ways in which it does.

However, in my own musings, I have stumbled into what might be an interesting exception: consciousness seems to play a role in behavior *when we think and talk about the mystery of consciousness.* When I contemplate "what it's like" to be something, that experience of consciousness presumably affects the subsequent processing taking place in my brain. And almost nothing I think or say when contemplating consciousness would make any sense coming from a system without it. How could an unconscious robot (or a philosophical zombie) contemplate conscious experience itself without having it in the first place? Imagine for a moment that David Chalmers himself is a zombie, completely lacking internal experience, and then consider the types of things he says in his book *The Conscious Mind* when explaining the concept of a zombie:

> Because my zombie twin lacks experiences, he is in a very different epistemic situation from me, and his judgments lack the corresponding justification. . . . I *know* I am conscious, and the knowledge is based solely on my immediate experience. . . . From the first-person point of view, my zombie twin and I are very different: I have experiences, and he does not.[9]

I don't see how a system that isn't conscious would ever have cause to produce these thoughts, let alone how an intelligent system would be able to make sense of them. Without ever having experienced consciousness, there's no *difference* that the Zombie Chalmers could be referring to. Chalmers's explanation for how a zombie is still conceivable in theory is that the language and concepts of consciousness could be built into the program of a zombie. A robot could certainly be programmed to describe specific processes like "seeing yellow" when it detects certain wavelengths of light, or even to talk about "feeling angry" under defined circumstances, without actually consciously seeing or feeling anything. But it seems impossible for a system to make a distinction between a conscious and unconscious experience in general without having an actual conscious experience as a reference point. When I talk about the mystery of consciousness—referring to something I can distinguish and wonder about and attribute (or not) to other entities—it seems highly unlikely that I would ever do this, let alone devote so much time to it, without feeling the experience I am referring to (for the qualitative experience is the entire subject, and without it, I can have no knowledge of it whatsoever). And when I turn these ideas over in my mind, the fact that my thoughts are *about the experience of consciousness* suggests that there is a feedback loop of sorts and that consciousness is affecting my brain processing. After all, my brain can think about consciousness only *after* experiencing it (one would presume).

Other than this one sinkhole I often fall into, however, most of our intuitions about what qualifies as evidence of consciousness affecting a system don't survive scrutiny. Therefore, we must reevaluate the assumptions we tend to make about the role consciousness plays in driving behavior, as these assumptions naturally lead to the conclusions we draw about what consciousness is and what causes it to arise in nature. Everything we hope to uncover through consciousness studies—from determining whether or not a given person is in a conscious state, to pinpointing where in the evolution of life consciousness first emerged, to understanding the exact physical process that gives birth to conscious experience—is informed by our intuitions about the function of consciousness.

5

When we talk about consciousness, we usually refer to a "self" that is the subject of everything we experience—all that we are aware of seems to be happening to or around this self. We have what feels like a unified experience, with events in the world unfolding to us in an integrated way. But, as we have seen, binding processes are partly responsible for this, presenting us with the illusion that physical occurrences are perfectly synchronized with our conscious experience of them in the present moment. Binding also helps solidify other percepts in time and space, such as the color, shape, and texture of an object—all of which are processed by the brain separately and melded together before arriving in our consciousness as a whole. Sometimes binding processes become interrupted, however, due to neurological disease or injury, leaving the sufferer in a confusing world where sights and sounds are no longer synced (disjunctive agnosia), or familiar objects are seen for their parts but are unrecognizable (visual agnosia).

Even with a healthy brain, we can sometimes catch small glitches in binding that shed light on the illusion it normally creates for us. A few months ago, I was walking to get a glass of water in the middle of the night when I heard a loud crash outside. For some reason, perhaps having to do with the fact that I was half-asleep, I experienced the moment in an unusual way: I noticed my body's startle response *before* I heard the sound of the crash. For a brief instant, I felt myself responding to something that "I" had not yet heard.

Imagine what your experience would be like if binding didn't take place at all—if, when playing the piano, for instance, you first saw your finger hit the key, then heard the note, and later finally felt the key hammer down. Or imagine if the process of binding were tampered with and you found yourself running *before* you heard the barking of the ferocious dog. Without binding processes, you might not even feel yourself to be a self at all. Your consciousness would be more like a flow of experiences in a particular location in space—which would be much closer to the truth. Is it possible to simply be aware of events, actions, feelings, thoughts, and sounds—all coming in a stream of awareness? Such an experience is not uncommon in meditation, and many people, myself included, can attest to it. The self we seem to inhabit most (if not all) of the time—a localized, unchanging, solid center of consciousness—is an illusion that can be short-circuited, without changing our experience of the world in any other way. We can have a full awareness of the usual sights, sounds, feelings, and thoughts, absent the sense of being a self who is the receiver of the sounds and the thinker of the thoughts. This is not at all at odds with modern neuroscience: an area of the brain known as the default mode network, which scientists believe contributes to our sense of self, has been found to be suppressed during meditation.[1]

There are other ways to suspend the sense of self. Psychedelic drugs—such as LSD, ketamine, and psilocybin—are known to quiet a circuit in the brain that connects the parahippocampus and the retrosplenial cortex in the default mode network, which explains why people describe losing their sense of self while under their influence.[2] Scientists study the experiences people have on psychedelic drugs and their related brain activity through fMRI (functional MRI scans). While under the influence of these drugs, participants report experiences ranging "from floating and finding inner peace, to distortions in time and a conviction that the self [is] disintegrating."[3] Many people assume that consciousness and the experience of self go hand in hand, but it is clear that in those moments when people report dropping the self, consciousness remains fully present. As Michael Pollan explains in his book *How to Change Your Mind*, on the scientific research of psychedelics:

> The more precipitous the drop-off in blood flow and oxygen consumption in the default mode network, the more likely a volunteer was to report the loss of a sense of self. . . . The psychedelic experience of "non-duality" suggests that consciousness survives the disappearance of the self, that it is not so indispensable as we—and it—like to think.[4]

Psychedelics also quell the communication among neurons in other areas beyond the default mode network, making activity in the brain less segregated in general. Erin Brodwin, a science journalist, discusses the work of Robin Carhart-Harris, who conducts imaging studies at Imperial College London on the impact of LSD on the brain:

> "The separateness of these networks breaks down, and instead you see a more integrated or unified brain," Carhart-Harris said. That change might help explain why the drug [LSD] produces an altered state of consciousness, too. . . . The barriers between the sense of self and the feeling of interconnection with one's environment appear to dissolve.[5]

Interestingly, one of the reasons people who take psychedelics inhabit such altered states is that this class of drug can also interrupt binding processes. It seems likely that this, too, contributes to a suspension of the feeling of being a self, distinct and separate from the world. Pollan points out that "our sense of individuality and separateness hinges on a bounded self and a clear demarcation between subject and object. But all that may be a mental construction, a kind of illusion."[6] Brodwin describes the experience of a participant in a study at Johns Hopkins on the therapeutic effects of psilocybin for patients with cancer and associated anxiety: "For a few hours, he remembers feeling at ease; he was simultaneously comfortable, curious, and alert. . . .

More than anything else, though, he no longer felt alone. 'The whole "you" thing just kinda drops out into a more timeless, more formless presence,' [he] said."[7]

Though it may be impossible for someone who hasn't experienced something like this to imagine it, consciousness can still persist without an experience of being a self, and even in the absence of thought. The journalist and author Michael Harris points out that it is partly because of this ability to interfere with one's sense of self that we know it is a construction:

> If the distinctness of the bodily self can be tampered with via such mechanical means [i.e., psychedelic drugs, a stroke, or a neurological disorder], then we must begin to accept that the bodily self—that feeling we are whole, inviolate beings—is not due to some special soul, or "I," resident behind our eyes.[8]

As mentioned, the typical notion of "self," along with other misperceptions of everyday experiences, can be overcome through meditation training, which is also now better understood at the level of the brain. For thousands of years, Eastern contemplative traditions have used meditation as an experimental basis for studying the nature of consciousness, and although Western science is a relative latecomer to these methods of introspection, research is now being conducted by neuroscientists on the specific effects of meditation on the mind and brain. This research will hopefully lead to new discoveries about how training our attention in systematic ways can provide a better understanding of consciousness and human psychology. At the very least, it confirms that valuable insights can be had through first-person tools of investigation. The Buddhist scholar Andrew Olendzki describes the illusory nature of self that can be revealed through meditation:

> Like the flatness of the earth or the solidity of the table, it [the notion of self] has utility at a certain level of scale—socially, linguistically, legally—but thoroughly breaks down when examined with closer scrutiny.[9]

Regardless of whether or not one can break through the illusion of self, however, there is obviously a wide range of what is being perceived in any given conscious experience—from someone in a minimally conscious state to someone piloting an aircraft. One thing we can confidently state, no matter what is being perceived, is that either consciousness is present or it isn't. It's either like something or it's not.

Just as we contemplated the moment at which conscious experience first appears in a developing embryo, we can wonder about the final moments of consciousness at the end of life. A friend of mine recently told me about spending time with his grandfather, who was slowly dying of heart disease. He described his grandfather's deterioration over the course of many months and the devastating experience of witnessing someone he knew well and loved dearly change so significantly. First to disappear was his grandfather's emotional regulation and impulse control, probably owing to damage occurring in his prefrontal cortex. His grandfather could no longer conceal his vacillating emotions, and everything he experienced—joy, frustration, lust, rage—was suddenly made known to everyone in the room. Next, his grandfather's memory began to fail, making the continuity of his personality less stable. Eventually, he lost the ability to speak and walk. At some point, my friend found himself wondering, as so many do in such situations, when his grandfather would truly no longer "be there." When would his grandfather cease to be "himself," and beyond that, when would his consciousness completely fade away? Sitting silently in a room without a recognizable personality, and with most of his memories gone, his grandfather still seemed to my friend to be experiencing *something.* Even when only the slightest glow of awareness remains, consciousness is obviously present in some form, up until the last moment it exists. And this minimal level of awareness—whatever it's like right before the lights go out altogether—may be completely unlike our familiar, human experience.

When Thomas Nagel asks us to imagine what it's like to be a bat, he is pointing out that we already know there are modes of consciousness vastly different from our own. Flying through space using echolocation must feel very different from walking down the sidewalk using vision. And the related, mind-boggling study of sensory substitution—whereby scientists have been able to give blind and deaf people new methods for perceiving what most of us see and hear—provides evidence that there is in fact a wide range of potential experiences in a brain. For example, with a tool called the BrainPort—a small grid that sits on the tongue and converts a video feed into minuscule electric shocks—the brain can begin to learn to interpret electro-tactile signals to the tongue. Using this technology, blind people can eventually accomplish tasks such as accurately throwing a ball into a basket and navigating an obstacle course.[10] Using a BrainPort is obviously related to using vision to maneuver around the physical world, but what the actual experience is like must be very different from seeing with one's eyes. There is a wonderful term, *umwelt*, introduced by the biologist Jakob von Uexküll in 1909, to describe the given experience of any particular animal, based on the senses used by that organism to navigate its environment. Bats have one umwelt, bees experience another, humans another, and someone using a technology like the BrainPort experiences yet another.

David Eagleman is involved in research that explores the possibilities of expanding our human umwelt to include information we don't currently have access to through our five senses. He explains that the brain "doesn't care how it gets the information, as long as it gets it."[11] At a 2015 TED conference, Eagleman described the potential future results of sensory substitution, whereby "new senses" are created for people:

> There's really no end to the possibilities on the horizon for human expansion. Just imagine an astronaut being able to feel the overall health of the International Space Station, or, for that matter, having you feel the invisible states of your own health, like your blood sugar and the state of your microbiome, or having 360-degree vision or seeing in infrared or ultraviolet.[12]

In fact, we know that the human brain, under the right conditions, can seamlessly integrate foreign objects into its map of what constitutes its body. The rubber-hand illusion is an example of how, when certain conditions are met, an outside object can become included in one's conception of self. In the original experiment, the subject sits with his real hand underneath a table, while a rubber hand rests on the table in its place. When the experimenter strokes the subject's real hand and the rubber hand simultaneously with a brush, the subject begins to feel that the rubber hand he sees on the table belongs to him. Later versions of the rubber-hand illusion have been demonstrated with the use of virtual reality. In one of these experiments, conducted by the neuroscientist Anil Seth and his team at the University of Sussex, the subject wears virtual reality goggles and experiences a virtual world in which she has a virtual hand. Sometimes the experimenters cause the hand to flash red in sync with the subject's heartbeat and sometimes it is out of sync. As we would expect, the subject has a greater feeling of ownership of the virtual hand when the flashing is in sync with her heartbeat.[13] Seth refers to our experiences of ourselves in the world as a kind of "controlled hallucination." He describes the brain as a "prediction engine" and explains that "what we perceive is its best guess of what's out there in the world." In a sense, he says, "we predict ourselves into existence."[14]

The "split-brain" phenomenon is also informative here, shedding light on both the malleability of consciousness and the concept of the self. Many people are now aware of the fascinating research conducted by Roger Sperry and Michael Gazzaniga at Caltech, beginning in the 1960s, on epilepsy patients who had undergone a corpus callosotomy. This is a surgical procedure in which the corpus callosum is cut, either partially or fully, separating connections between the left and right hemispheres of the brain in an effort to prevent seizures from spreading. Although these split-brain patients appeared surprisingly unchanged by the procedure, research on them revealed a bizarre and counterintuitive reality that calls into question many of our assumptions about the fluidity and boundaries of consciousness.

In experiments on people who have undergone split-brain surgery, information can be given separately to each of their two brain hemispheres through vision (in the form of pictures, written language, etc.), because the right visual field is projected to the left hemisphere of the brain and vice versa. In a normal person, the information coming through either visual field is shared with the opposite hemisphere of the brain through the corpus callosum. In split-brain patients, visual stimulus to each field is received by only one side of the brain. The same goes for stimuli presented to each ear, as well as for most of the information from patients' hands—for the most part, the touch receptors from each hand project to the opposite hemisphere of the brain, and the movement of each hand is also controlled by the opposite hemisphere. In fact, after surgery, split-brain patients can experience something called "hemispheric rivalry," in which they are seen attempting opposing behaviors with their left and right hands in a disconcerting battle—such as trying to button up their shirt with one hand while the other hand is busy at work unbuttoning; attempting to hug a spouse with one arm while pushing him away with the other; and simultaneously opening and closing a door with opposite hands.[15]

Neuroscientists have developed a variety of creative methods for receiving communications from the two hemispheres of split-brain patients as well, revealing other startling aspects of the condition. In the vast majority of people, the left hemisphere is responsible for the expression of language through speech and writing, leaving the right hemisphere mute; however, the right hemisphere is able to communicate through nodding and gestures of the left hand (and singing, in some cases).[16] If a subject is given a coin to hold in her left hand without being able to see it, only the right hemisphere will be aware of it. When asked what she is holding, she will respond that she has no idea, because the left hemisphere (which maintains the ability to communicate verbally) has no awareness of the coin. But if asked to point to a picture of the object she was given, her left hand (controlled by her right hemisphere, the one that knows about the coin) will correctly point to the picture of a coin. Similarly, if the word "key" is presented to a subject's left visual field and he is asked what word he sees, he will report that he doesn't see anything—his speaking, left hemisphere can't see the word. Yet if he's asked to pick up the object corresponding to the word that is up on the screen, he will reach out with his left hand (controlled by the right hemisphere, which sees the word) and pick up the key (Figure 5.1). This type of experiment can be repeated in a variety of ways, producing the same results time and again. In fact, split-brain patients sometimes report (via the speaking left hemisphere) that their left hand acts on its own—closing the book they are reading, for instance—a confirmation that "they" are unaware of the desires and intentions of the right hemisphere.

To the surprise of the first neuroscientists to conduct such experiments (and to the rest of us!), it seems that the same person can have two different answers to a question, along with completely different desires and opinions in general. And even more astonishing is the discovery that the feelings and opinions of each hemisphere seem to be privately experienced and unknown to the other. One

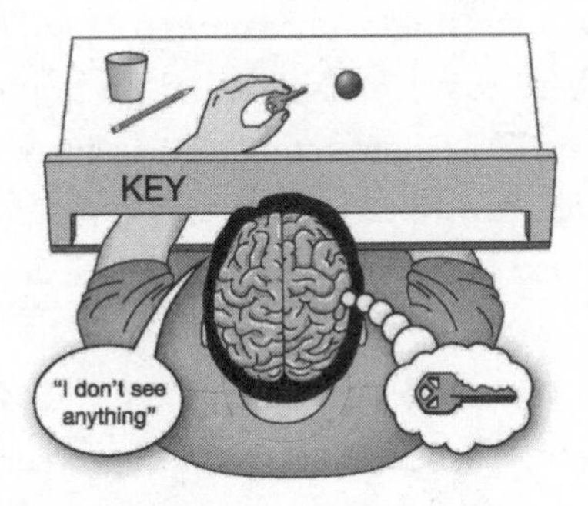

Figure 5.1: Split-brain study.

"self" of a split-brain patient is as puzzled by the opinions and desires of the other as another person in the room would be. Whether or not both points of view in split-brain patients are conscious is difficult if not impossible to answer, but we have no reason to doubt that there is an experience associated with the thoughts and desires of each, and most neuroscientists believe that both hemispheres are in fact conscious. As the neuroscientist Christof Koch, of the Allen Institute for Brain Science, points out, "Because both the speaking *and* the mute hemispheres carry out complex, planned behaviors, both hemispheres will have conscious percepts, even though the character and content of their feelings may not be the same."[17]

The split-brain literature contains many examples suggesting that two conscious points of view can reside in a single brain. Most of them also topple the typical notion of free will, by exposing a phenomenon generated by the left hemisphere that Gazzaniga and his colleague Joseph LeDoux dubbed "the interpreter."[18] This phenomenon occurs when the right hemisphere takes action based on information it has access to that the left hemisphere doesn't, and the left hemisphere then gives an instantaneous and false explanation for the split-brain subject's behavior. For example, when the right hemisphere is given the instruction "Take a walk" in an experiment, the subject will stand up and begin walking. But when asked why he's leaving the room, he will give an explanation such as, "Oh, I need to get a drink." His left hemisphere, the one responsible for speech, is unaware of the command the right side received, and we have every reason to think that he does in fact believe his thirst was the reason he got up and began walking. As in the example in which experimenters were able to cause a feeling of will in subjects who in actuality were not in control of their own actions, the phenomenon of "the interpreter" is further confirmation that the feeling we have of executing consciously willed actions, at least in some instances, is sheer illusion.

Regardless of what the split-brain research tells us about conscious will, however, the more basic insight is more relevant to our discussion: different sets of intentions in a split-brain patient seem to be relegated to distinct and separate islands of consciousness. In the example of a patient's battle with herself over buttoning up a shirt, one side feels that her right hand is being controlled by "someone else," who is fighting against her action to put on a chosen shirt.

The other side is rejecting a bad wardrobe choice that "someone else" made. In moments such as these, a split-brain patient behaves (and probably feels) more like two conjoined twins than like a single person.

In his book *The Master and His Emissary*, about the two hemispheres of the brain, the psychiatrist Iain McGilchrist describes his intriguing thesis about the possibility that consciousness originates much deeper in the structures of the brain than scientists typically believe:

> It seems to me more fruitful to think of consciousness not as something with sharp edges that is suddenly arrived at once one reaches the very top of mental functioning, but as a process that is gradual, rather than all-or-nothing, and begins low down in the brain. . . . The problem then becomes not how two wills can *become* one unified consciousness, but how one field of consciousness can accommodate two wills. . . . Consciousness is not a bird, as it often seems to be in the literature—hovering, detached, coming in at the top level and alighting on the brain somewhere in the frontal lobes—but a tree, its roots deep inside us.[19]

With the revelations brought to us through split-brain research and other advances in modern neuroscience, many have been led to the following question: Is there some version of split consciousness that occurs in brains that aren't physically split? Are there other centers of consciousness, even what we might think of as other minds, residing closer to us than we think? Perhaps it's not impossible to imagine that different "centers," "configurations," or "flows" of consciousness exist in close proximity to one another or overlap, even in a single human body.

6

IS CONSCIOUSNESS EVERYWHERE?

We seem to be left without answers to the two questions with which we began this investigation: when we look closely, we can't find reliable external evidence of consciousness, nor can we conclusively point to any specific function it serves. These are both deeply counterintuitive outcomes, and this is where the mystery of consciousness starts bumping up against other mysteries of the universe.

If we can't point to anything that distinguishes which collections of atoms in the universe are conscious from those that aren't, where can we possibly hope to draw the line? Perhaps a more interesting question is why we should draw a line at all. When we view our own experience of consciousness as being "along for the ride," we suddenly find it easier to imagine that other systems are accompanied by consciousness as well. It's at this point that we must consider the possibility that *all* matter is imbued with consciousness in some sense—a view referred to as panpsychism.[1] If the various behaviors of animals can be accompanied by consciousness, why not the reaction of plants to light—or the spin of electrons, for that matter? Perhaps consciousness is embedded in matter itself, as a fundamental property of the universe. It *sounds* crazy, but as we will see, it's worth posing the question.

The term *panpsychism*, coined by the Italian philosopher Francesco Patrizi in the sixteenth century, is derived from the Greek *pan* ("all") and *psyche* ("mind" or "spirit"). Some versions of panpsychism describe consciousness as separate from matter and composed of some other substance, a definition reminiscent of vitalism and traditional religious descriptions of a soul. But while the term has been used to describe a wide range of thinking throughout history, contemporary considerations of panpsychism provide descriptions of reality very different from the earlier versions—and are unencumbered by any religious beliefs.

One branch of modern panpsychism proposes that consciousness is intrinsic to all forms of information processing, even inanimate forms such as technological devices; another goes so far as to suggest that consciousness stands alongside the other fundamental forces and fields that physics has revealed to us—like gravity, electromagnetism, and the strong and weak nuclear forces. The full range of serious deliberations regarding panpsychism—whether they narrow in on certain types of information processing or apply to all matter universally—are unlike most of the panpsychic theories of the past. Modern thinking about panpsychism is informed by the sciences and is fully aligned with physicalism and scientific reasoning.

I love the title of an article by the philosopher Philip Goff: "Panpsychism Is Crazy, but It's Also Most Probably True." His line of thinking follows this path:

> Once we realise that physics tells us nothing about the intrinsic nature of the entities it talks about, and indeed that the only thing we know for certain about the intrinsic nature of matter is that at least some material things have experience . . . the theoretical imperative to form as simple and unified a view as is consistent with the data leads us quite straightforwardly in the direction of panpsychism.[2]

It's because of the value of simplicity that I tend to favor the branch of panpsychism that describes consciousness as fundamental to matter—as opposed to requiring a certain level of information processing for consciousness to exist. This, once again, is a result of the hard problem of consciousness, which crops up anywhere you attempt to draw a line—whether at neuronal processing or at simpler forms of information processing. Although in many ways it's more difficult to get our minds around, the view that consciousness is intrinsic to matter is a more convincing solution to me, in part because it is a simpler one (albeit only slightly). Consider the Higgs field as an analogy: Physicists knew that the Higgs field had to exist—if it didn't, the electrons and quarks that make up all of us would be massless and travel at the speed of light. For years before the discovery of its carrier, the Higgs boson, they posited a Higgs field. Although nothing about its confirmation supports (or provides any evidence for) theories about consciousness, it helps us understand the analogous proposition in panpsychism—that perhaps consciousness is another property of matter, or of the universe itself, that we have yet to discover.

In his book *Panpsychism in the West*, the philosopher David Skrbina provides a survey of the history of scientific arguments for panpsychism that are based on rationalism, empirical evidence, and evolutionary principles. After Darwin's theory of evolution by natural selection was published (1859) and subsequent advances in the fields of physics, chemistry, and biology revealed that human beings were composed of the same elements as other matter, the true mystery of consciousness became apparent. And the new understanding that everything in the universe consisted of the same building blocks led to further support for a scientific and evolutionary perspective entailing some form of panpsychism. The natural tendency of scientific exploration is to arrive at as simple an explanation as possible, and the concept of consciousness emerging out of nonconscious material represents a kind of failure of the typical goal of scientific explanation. In philosophy, this jump from a nonconscious to a conscious state of matter is referred to as "radical" or "strong" emergence.[3] Skrbina quotes the celebrated biologist J. B. S. Haldane on his opposition to the notion of radical emergence, based on the inevitable complexity it adds to any explanation of consciousness:

> If consciousness were not present in matter, this would imply a theory of strong emergence that is fundamentally anti-scientific. Such emergence is "radically opposed to the spirit of science, which has always attempted to explain the complex in terms of the simple. . . . If the scientific point of view is correct, we shall ultimately find them [signs of consciousness in inert matter], at least in rudimentary form, all through the universe."[4]

Skrbina walks the reader through more than three hundred years of contemplations by scientists—from Johannes Kepler to Roger Penrose—who take a scientific approach to panpsychism, many of whom arrive at the conclusion that the simplest explanation of consciousness is in fact a panpsychic one. About thirty years after Haldane, in the 1960s, the biologist Bernhard Rensch asserted that just as there is a blurring of categories when we examine the evolution of one life-form to another at the level of microorganisms and cells, the stark division between living and nonliving systems is blurred, and a mistaken distinction likely carries over to the boundaries of conscious experience as well.[5]

Additionally, when scientists assume they have bypassed the hard problem by describing consciousness as an emergent property—that is, a complex phenomenon not predicted by the constituent parts—they are changing the subject. All emergent phenomena—like ant colonies, snowflakes, and waves—are still descriptions of matter and how it behaves as witnessed from the outside.[6] What a collection of matter is like from the *inside* and whether or not there is an experience associated with it is something the term "emergence" doesn't cover. Calling consciousness an emergent phenomenon doesn't actually explain anything, because to the observer, matter is behaving as it always does. If some matter has experience and some doesn't (and some emergent phenomena entail experience and some don't), the concept of emergence as it is traditionally used in science simply doesn't explain consciousness.

Figure 6.1: Emergence. A phenomenon that is not predicted by the constituent parts, and is more complex than the sum of its parts, is referred to as an emergent phenomenon.

Some philosophers go so far as to suggest that there isn't a hard problem of consciousness at all, reducing consciousness to an illusion. But as others have pointed out, consciousness is the one thing that can't be an illusion—by definition. An illusion can appear *within* consciousness, but you are either experiencing something or you're not—consciousness is necessary for an illusion to take place. In his essay "The Consciousness Deniers," the British analytic philosopher Galen Strawson analyzes this view of consciousness-as-illusion and expresses exasperation with the utter incoherence of the idea: "How could anybody have been led to something so silly as to deny the existence of conscious experience, the only general thing we know for certain exists?"[7] The philosopher Ned Block, of NYU's Center for Neural Science, describes something he's observed in his students akin to different personality types when he lectures about the hard problem of consciousness. He estimates that about one-third of his students "don't appreciate phenomenology [felt experience] and the difficult problems it raises," and he thinks it would be interesting to study the neurological difference between people who are able to intuitively grasp the hard problem and those who aren't (or who view it as an illusion).[8] Regardless, relegating consciousness to the status of an illusion misses the point, in my view. In effect, it is simply redefining consciousness as "the illusion of consciousness." Even if we agreed to call consciousness an illusion, which seems absurd, we would still wonder how deep this illusion goes. Are other complex processes, or other collections of matter, experiencing this "illusion"? All the questions of consciousness and panpsychism would still stand before us.[9]

In fact, Strawson posits that "panpsychism is the most plausible theoretical view to adopt if one is an out-and-out naturalist . . . who holds that physicalism is true," that "everything that concretely exists is physical," and that "all physical phenomena are forms of energy." He concludes that "panpsychism is simply a hypothesis about the ultimate intrinsic nature of this energy, the hypothesis that the intrinsic nature of energy is experience. . . . Physics is untouched by this hypothesis. Everything true in physics remains true."[10]

Nevertheless, scientific considerations of panpsychism are still seen as controversial and are contrary to the conventional scientific view. While consciousness is notoriously difficult to study and even to define, most neuroscientists believe that it results from complex processes in the brain, and that we'll eventually discover the ultimate cause of consciousness by studying its neural correlates. Many neuroscientists do admit, however, that the hard problem will persist, because scientific understanding, no matter how complete, seems to have no way of offering us direct insight into the subjective experience associated with those physical properties—studying systems like the brain simply delivers us more information about physical properties. The neuroscientist V. S. Ramachandran, for example, has conceded that "qualia" (the experiential qualities of consciousness that we can label, such as what it's like to see the color blue or feel something sharp) will remain a puzzle:

> Qualia are vexing to philosophers and scientists alike because even though they are palpably real and seem to lie at the very core of mental experience, physical and computational theories about brain function are utterly silent on the question of how they might arise or why they might exist.[11]

Neuroscientists who study consciousness are most interested in the differences at the level of the brain between seemingly conscious and unconscious *functions* of the body (you're aware of reading the words on this page at this moment, but you're not aware of the activities of your kidneys) and conscious and unconscious *states* (being awake versus being in deep sleep, for example). There are a variety of hypotheses proposing that certain areas of the brain, or types of neural processing, create a conscious experience; some scientists, Francis Crick and Christof Koch among them, have even speculated that it is the frequency at which neurons fire that causes them to give rise to consciousness.[12]

Crick and Koch attempted to pinpoint the source of consciousness in the brain by conducting research on the visual system. They hoped to better understand which types of visual stimuli we process consciously (are aware of seeing), which stimuli the brain is responding to but we have no conscious awareness of (subliminal processing), and which areas of the brain are responsible for these different kinds of processing. While useful and interesting, this type of research is, once again, limited. It increases our knowledge of the brain and our human experience, but it can't tell us anything about what *consciousness is in the first place*, nor does it help us understand whether or not other types of systems, animate or inanimate, could be experiencing it.

More recently, the neuroscientist Giulio Tononi, director of the University of Wisconsin–Madison's Center for Sleep and Consciousness, together with Marcello Massimini and his team at the University of Milan, formulated what may become a method for determining whether or not a person is conscious. In the procedure, nicknamed "zap and zip," transcranial magnetic stimulation (TMS) is used to deliver a pulse of magnetic energy to the brain, and the activity of the subsequent electric current running through the cortical neurons is then read by EEG.[13] The resulting patterns are mapped onto a "perturbational complexity index" (PCI). Koch explains that the method establishes a PCI cutoff value "as a critical threshold—the minimum measure of complex brain activity—supporting consciousness."[14] This method attempts to detect consciousness in subjects whose level of awareness is hard to decipher from external cues—including subjects in deep sleep, anesthetized subjects, and patients in a coma. It hopefully places us one step closer to determining whether brain-damaged, locked-in, or late-stage dementia patients are in a minimally conscious state versus a "vegetative state," or whether a surgical patient has become conscious while under anesthesia—conditions we currently have only limited tools for detecting.

This is admittedly some of the most important work being done in neuroscience today, but again, questions regarding conscious versus unconscious functions or states of the brain do not necessarily address the larger questions regarding *what consciousness is* and how deep in the universe it runs. The fact remains, however, that the majority of scientists believe that consciousness is an emergent phenomenon resulting from neuronal processing. Most assume that if "we" are not conscious of certain experiences and brain processes, there must not be any experience associated with them at all. This may be true, but as we will see, it may not make sense to follow that line of reasoning.

Let's inspect how this research informs (or fails to inform) panpsychic views. There are inconsistencies in many of the hypotheses put forth by scientists and philosophers. They show up in:

1. attempts to draw a line between where we are likely to find consciousness and where we are not—this usually has something to do with information processing; and
2. the failure of scientists and philosophers to overcome the strong, though probably false,

intuition that there can't be more than one center or system of consciousness residing in a human body.

Christof Koch is one neuroscientist who is willing to consider the panpsychic interpretation, telling an interviewer:

> If you take a more conceptual approach to consciousness, the evidence suggests there are many more systems that have consciousness—possibly all animals, all unicellular bacteria, and at some level maybe even individual cells that have an autonomous existence. We might be surrounded by consciousness everywhere and find it in places where we don't expect it because our intuition says we'll only see it in people and maybe monkeys and also dogs and cats. But we know our intuition is fallible, which is why we need science to tell us what the actual state of the universe is.[15]

I follow him here wholeheartedly, but he then goes on to say things like, "We know that most organs in your body do not give rise to consciousness. Your liver, for example, is very complicated, but it doesn't seem to have any feelings."[16] If one can imagine that a worm has some level of consciousness (and that it would maintain its consciousness while residing in a human body), whether it's contributing to the scope of consciousness that "I" am experiencing right now is irrelevant to the question of whether the worm is experiencing something. So these separate lines of investigation (what contributes to "my" consciousness versus what *is conscious*) end up confusing the larger question about what consciousness is in the first place and where in the universe we will find it.

By entertaining the notion that bacteria or individual cells could have some level of consciousness, Koch seems open to a modern version of panpsychism, yet in the same conversation he asserts that the cerebellum, with its sixty-nine billion neurons, "does not give rise to consciousness." But just because the cerebellum is not responsible for the part of my brain that governs language or for the flow of consciousness that I consider to be "me," we can still wonder whether it's *another region* (or regions) of consciousness, just as we can speculate that a worm or a bacterium might be conscious. Although Koch is here addressing consciousness in two different contexts—considering a panpsychic view in one instance, and pointing to specific processes in the body that aren't included in the typical experience of consciousness in the other—the overall thinking on this subject in neuroscience and philosophy tends to be inconsistent; or at the very least, a piece of the conversation is often missing.

And even though, as mentioned earlier, Thomas Nagel's definition of the word "consciousness" (i.e., being *like something*) is the most accurate way to talk about subjective experience, there are a variety of ways people use the word (the capacity for self-reflection, wakefulness, alertness, etc.), which causes additional confusion. But we can continue to pose questions about whether consciousness exists outside systems that can report back about it—we just have to do so on another level of conversation. When I am unconscious during a period of deep sleep, for instance, all we know is that the part of the system that makes up "me" has been interrupted; the continuity (and even the reality) of my experience ceases for a period because the operation of that part of the system ceases. But whether consciousness itself carries on in other areas of my brain or body while the experience of "me" is on hold is still an open question.

No matter how much knowledge we gain about the workings of the brain, the question at hand is likely to remain unanswered: How deep in the universe does consciousness run? In *The Conscious Mind*, David Chalmers suggests that consciousness could be manifested in the functioning of something as basic as a simple technological device:

> As we move along the scale from fish and slugs through simple neural networks all the way to thermostats, where should consciousness wink out? . . . The thermostat seems to realize the sort of information processing in a fish or a slug stripped down to its simplest form, so perhaps it might also have the corresponding sort of phenomenology in its most stripped-down form. It makes one or two relevant distinctions on which action depends; to me, at least,

it does not seem unreasonable that there might be associated distinctions in experience.[17]

So if it's plausible that worms or bacteria (or thermostats!) are accompanied by some level of consciousness, however minimal and unlike our own experience, why not follow the same logic when it comes to organs in the body, or the cerebellum (which contains most of the neurons in the brain)? Just because something isn't appearing in the field of what "I" am experiencing, why rule out the possibility that many forms of consciousness exist simultaneously within the boundaries of my body?

Another potential source of erroneous arguments against panpsychism is based in evolution, as most scientific and philosophical support for the idea that consciousness is confined to the nervous systems of living things relies in part on the assertion that consciousness is a product of biological evolution. The logic is understandable, given that our most sophisticated methods of survival seem to us to require consciousness. But if consciousness doesn't determine our behavior as we have traditionally assumed, the evolution argument doesn't hold up. How can consciousness increase the likelihood of survival if it doesn't affect our behavior in the typical sense?

When we look outside the context of animal life, where it's easier for us to drop our ingrained intuitions, we find that it's actually hard to intuit the logic that any amount of information processing, no matter how complex, would suddenly cause *those processes* to become conscious. When your golden retriever runs to greet you at the end of the day, her consciousness seems as obvious to you as any other fact. But as we've seen, even when we imagine robots that look and act like human beings, we seem to be unable to determine whether or not they would be conscious. It's only because we experience consciousness so readily, and ascribe it to other life-forms by analogy so easily, that it seems like an obvious capacity (and that we're not continually shocked to be experiencing something in every waking moment).[18] We should be as surprised by the reality of our own consciousness as we would be to learn that the latest smartphone is conscious.

My own sense of the correct resolution to the mystery of consciousness, whether or not we can ever achieve a true understanding, is still currently split between a brain-based explanation and a panpsychic one. But while I'm not convinced that panpsychism offers the correct answer, I *am* convinced that it is a valid category of possible solutions that cannot be as easily dismissed as many people seem to think. Unfortunately, it remains difficult for scientists to join the conversation without fear of jeopardizing their credibility. In a 2017 essay titled "Minding Matter," Adam Frank, a professor of astrophysics at the University of Rochester, eloquently expresses both the mystery of consciousness and the reluctance of scientists to propose theories that venture beyond viewing consciousness as a result of processing in the brain:

> It is as simple as it is undeniable: after more than a century of profound explorations into the subatomic world, our best theory for *how matter behaves* still tells us very little about *what matter is*. Materialists appeal to physics to explain the mind, but in modern physics the particles that make up a brain remain, in many ways, as mysterious as consciousness itself. . . . Rather than trying to sweep away the mystery of mind by attributing it to the mechanisms of matter, we must grapple with the intertwined nature of the two. . . . Consciousness might, for example, be an example of the emergence of a new entity in the Universe not contained in the laws of particles. There is also the more radical possibility that some rudimentary form of consciousness must be added to the list of things, such as mass or electric charge, that the world is built of.[19]

But while theoretical physicists can happily propose ideas such as the predictions of string theory—from ten (or more) dimensions of space to the vast landscape of possible universes—and still have their work get a fair hearing, it is considered a risk to one's reputation to suggest that consciousness might exist outside the brain.

Frank points out a similar double standard that is applied to evaluating the various interpretations of quantum mechanics: "Why does the infinity of parallel universes in the many-worlds interpretation get associated with the sober, hard-nosed position, while including the perceiving subject [consciousness] gets condemned as crossing over to the shores of anti-science at best, or mysticism at worst?"

Although some scientists have been led naturally to a panpsychic view in one form or another, the term still carries the stink of the New Age. David Skrbina explains that, at first mention, the idea that the inanimate world possesses consciousness seems so anti-scientific that it incites a reflexive and concerted opposition:

> Upon laying out a panpsychist position, one is immediately faced with the charge that he believes that "rocks are conscious"—a statement taken as so obviously ludicrous that panpsychism can be safely dismissed out of hand. . . . We may see strong analogies with the human mind in certain animals, and so we apply the concept [consciousness] to them with varying degrees of confidence. We may see no such analogies to plants or inanimate objects, and so to attribute consciousness to them seems ridiculous. This is our human bias. To overcome this anthropocentric perspective, the panpsychist asks us to see the "mentality" of other objects not in terms of *human* consciousness but as a subset of a certain *universal quality* of

> physical things, in which both inanimate mentality and human consciousness are taken as particular manifestations.[20]

Though all the attacks on panpsychism I've read lack substantive, detailed arguments, they have been fierce. From the *Encyclopedia of Philosophy* (Edwards, 1967) to the *New York Review of Books*, panpsychism has been accused of being "unintelligible" and "breathtakingly implausible," with its adherents likened to "religious fanatics."[21]

Those of us who want to push this conversation forward have an important obligation to clearly distinguish panpsychic views from the false conclusions people tend to draw from them—namely, that panpsychism somehow justifies or explains a variety of psychic phenomena—following from the incorrect assumption that consciousness must entail a mind with a single point of view and complex thoughts. Ascribing some level of consciousness to plants or inanimate matter is not the same as ascribing to them *human* minds with wishes and intentions like our own. Anyone who believes the universe has a plan for us or that he can consult with his "higher self" for medical advice should not feel propped up by the modern view of panpsychism. It supports nothing of the sort. Bacteria with some minimal level of consciousness streaming through their atoms would still be *bacteria*. They would still lack brains and complex minds, much less human ones.

As the philosopher Gregg Rosenberg points out, when

we entertain the notion that a bacterium or an atom possesses some level of conscious experience, "we are obviously not attributing to it the qualities of our own experiences," but instead we can imagine "a qualitative field that has a character in some very *abstract* sense like that of our experiences, but *specifically* unimaginable to us and unlike our own [experience of consciousness]."[22] And, of course, the false conclusions drawn from a misunderstanding of panpsychism—that individual atoms, cells, or plants possess an experience comparable to that of a human mind, for instance—are often the very thing used to argue against it. Unfortunately, it seems quite hard for us to drop the intuition that consciousness equals complex thought. But if consciousness is in fact a more basic aspect of the universe than previously believed, that doesn't suddenly give credence to your neighbor's belief that she can communicate telepathically with her ficus tree. In actuality, if a version of panpsychism is correct, everything will still appear to us and behave exactly as it already does.

7

BEYOND PANPSYCHISM

Imagine being a brain without any sense organs connected, floating in empty space or in a vast body of water. Then imagine your senses being connected, one at a time. First vision. The only content available to you is a subtle experience of sight. You can see light perhaps—pulsating light of varying brightness, coming in and out. Try to apprehend this without including the concepts of memory or language, so that there's no sense of a self thinking, *Whoa, it was just dark but now it's light again!* Instead, try to imagine a very simple flow of "first experiences": light and dark alternating, then brighter light, dimmer light, pulsating light. Next, imagine light that takes on shapes: a circular light, a beam of light, light that extends far into the distance. Adding color perhaps: a reddish light that transforms to orange, then yellow, then blue. Imagine feeling formless and weightless. You're free-floating in space, with no thoughts or concepts—no words "orange" or "red," just the pure *experience* of those colors. Visualize the most basic experience imaginable. Next, bring in sounds, then tastes, then smells—each arriving one at a time in as pure a form as possible. You're simply experiencing what arrives in your awareness, without words or concepts to describe what the experience is like. And finally, imagine the feeling of touch coming online in the form of pressure or heat—spanning broad areas and in small, pinpoint locations—not in your *body*, of course, since you don't have one, but in locations in space. . . .

It's difficult to maintain this type of imagery for long, but we can get enough of a sense of such a state to imagine that, at the very least, something like it is *possible*.

Most people who have had sufficient training in meditation realize that an experience of consciousness needn't be accompanied by thoughts—or any input to the senses, for that matter. It seems possible to be acutely aware of one's subjective experience in the absence of thought, sights, sounds, or any other perception. As we have seen, the feeling we have of being a concrete self, and the intuitions that come along with it, create formidable obstacles to thinking creatively about consciousness. These intuitions also contribute to our tendency to reflexively reject panpsychism as a plausible category of theories, even when so many arrows of logic point in its direction. But when we zoom in on the actual details, the idea looks less improbable. Rebecca Goldstein makes the case that we in fact already know that consciousness is integral to matter because we are made of matter ourselves, and it is the one property we have direct access to:

> Consciousness is an intrinsic property of matter; indeed, it's the only intrinsic property of matter that we know, for we know it directly, by ourselves being material conscious things. All of the other properties of matter have been discovered by way of mathematical physics, and this mathematical method of getting at the properties of matter means that only relational properties of matter are known, not intrinsic properties.[1]

Galen Strawson makes a similar point by turning the mystery of consciousness on its head. He argues that consciousness is in fact the only thing in the universe that is *not a mystery*—in the sense that it is the only thing we truly understand firsthand. According to Strawson, it is *matter* that's utterly mysterious, because we have no understanding of its intrinsic nature. And he has dubbed this "the hard problem of matter":

> [Physics] tells us a great many facts about the *mathematically describable structure* of physical reality, facts that it expresses with numbers and equations . . . but it doesn't tell us anything at all about the intrinsic nature of the stuff that fleshes out this structure. Physics is silent—perfectly and forever silent—on this question. . . . What is the fundamental stuff of physical reality, the stuff that is structured in the way physics reveals? The answer, again, is that we don't know—

> except insofar as this stuff takes the form of conscious experience.[2]

Once again, it's important to distinguish between consciousness and complex thought when considering modern panpsychic views. Postulating that consciousness is fundamental isn't the same as suggesting that complex ideas or thoughts are fundamental and magically result in a material realization of those ideas (a common misinterpretation of panpsychism). The claim is just the opposite—that if consciousness exists as a fundamental property, complex systems, built from that-which-is-already-streaming-consciousness, could eventually give rise to physical structures such as human minds. David Skrbina addresses the problem of anthropic projections, in which we "place the demands of human consciousness on inanimate particles," and he explains the necessity of distinguishing between consciousness and *memory*:

> Certainly anything like the human mind requires a human-like memory, but this is relevant only for complex organisms. It is not reasonable to demand that atomic particles have anything like the memory capability of the human being, or even any physical instantiation of something like memory. Minds of atoms may conceivably be, for example, a stream of instantaneous memory-less moments of experience.[3]

Though many people wonder: If the most basic constituents of matter have some level of conscious experience, how could it be that when they form a more complex physical object or system, those smaller points of consciousness combine to create a new, more complex sphere of consciousness? For instance, if all the individual atoms and cells in my brain are conscious, how do those separate spheres of consciousness merge to form the consciousness "I'm" experiencing? What's more, do all the smaller, individual points of consciousness cease to exist after giving birth to an entirely new point of consciousness? This is referred to as "the combination problem," which the *Stanford Encyclopedia of Philosophy* describes as "the hardest problem facing panpsychism," noting:

> The problem is that this is very difficult to make sense of: "little" conscious subjects of experience with their micro-experiences coming together to form a "big" conscious subject with its own experiences. . . . The idea of many minds forming some other mind is much harder to get your head around (so to speak).[4]

For many scientists and philosophers, the combination problem presents the biggest obstacle to accepting any description of reality that includes consciousness as a widespread feature. However, the obstacle we face here once again seems to be a case of confusing *consciousness* with the concept of a *self*, as philosophers and scientists tend to speak in terms of a "subject" of consciousness. The term "self" is usually used to describe a more complex set of psychological characteristics—including qualities such as self-confidence or a capacity for empathy—but a "subject" still describes an experience of self in its most basic form. In a paper discussing the problem that combination poses, David Chalmers writes, "How could any phenomenal relation holding between distinct subjects . . . suffice for the constitution of a wholly new subject?"[5] But perhaps it's wrong to talk about a subject of consciousness, and it's more accurate to instead talk about the *content available* to conscious experience at any given location in space-time, determined by the matter present there—umwelts applied not just to organisms, but to all matter, in every configuration and at every point in space-time.

Viewed in this way, the combination problem no longer seems to be an obstacle to all versions of panpsychism. Rather, it may be an additional reason to favor a perspective in which consciousness is a fundamental feature of the universe, as opposed to being confined to some level of information processing. Considering consciousness to be fundamental allows for matter to have a certain internal character everywhere, in all its different forms. And in this view, consciousness is not interacting *with itself*, as it would be in the act of "combining." This line of thinking yields interesting questions: Does certain content appear in an area of consciousness depending on the matter present in that location in space-time? Are there overlapping experiences, as well as merging experiences, of content?

In a recent conversation I had with Christof Koch, we discussed what might result from a hypothetical experiment in which two brains were connected together as successfully as the two hemispheres of an ordinary brain are connected. Since it seems as though the mind and the contents of consciousness can be divided in a split-brain patient, would two brains wired together produce a new, integrated mind? If Christof and I had our brains wired together, for instance, would it create a new Christof-Annaka consciousness—a new single point of view? Would a new mind be produced, with access to all the content that had previously been experienced separately by our brains—all our thoughts, memories, fears, abilities, and so forth—constituting a new "person"?[6] Even if the answer is yes, which it probably is, I don't think we encounter a combination problem in this thought experiment. We run into problems only if we see the conscious experiences of myself and Christof as selves or subjects—permanent structures of consciousness with fixed boundaries. In the instance of connecting two brains, we might simply have an example of consciousness changing its content or character—in the same way that the content of your consciousness changes when you close and open your eyes: the trees and sky are available to your field of view and then they're not. When you dream, you experience environments quite different from your actual surroundings, maybe even feeling yourself to be a different person altogether. And in deep sleep, you lose consciousness entirely, only to gain it back again. During both of my pregnancies, I found myself experiencing drastic variations in the contents of my consciousness—sensations in my uterus I had never before known were on the menu of experience, an obsession with tomatoes and tomato sauces in every form, feelings of panic and other, more amorphous emotional rides, physical pain, insomnia. . . . I didn't feel like "myself," and I expect I wouldn't feel like myself during a mind meld with a sixty-year-old male neuroscientist, either, but it doesn't necessarily point to a *combination problem* for consciousness. Even in our daily lives, content comes and goes, and consciousness itself can seem to flicker in and out.

We run into a combination problem only when we drag the concept of a "self" or a "subject" into the equation. But we know that the idea of the self, as a concrete entity, is an illusion. It's admittedly a very tough illusion to relinquish, but I think the solution to the combination problem is that there is really no "combining" going on at all with respect to consciousness itself. Consciousness could persist as is, while the character and content change, depending on the arrangement of the specific matter in question. Maybe content is sometimes shared across large, intricately connected regions and sometimes confined to very small ones, perhaps even overlapping. If two human brains were connected, both people might feel as if the content of their consciousness had simply expanded, with each person feeling a continuous transformation from the content of one person to the whole of the two, until the connection was more or less complete. It's only when you insert the concepts of "him," "her," "you," and "me" as discrete entities that the expanding of content for any area of consciousness (or even multiple areas merging) becomes a combination problem. It reminds me of the classic device of characters switching places in a story or film, giving them an experience of what it's like to be someone else. When we look closely at what this actually entails, it becomes impossible to even pose the question. Where's the "me" that would switch if I became someone else? Being someone else would be no different from what it's already like to be that person. It seems paradoxical, but we end up simply stating the obvious: "That's what it's like to be over there as that configuration of atoms, and this is what it's like to be over here as this configuration of atoms." It's analogous to saying, "The configuration of atoms that compose a leaf result in all its expected leaf properties, and a collection of H_2O molecules take on the expected properties of water. That's what molecules *do* in that configuration, and this is what they *do* in this configuration. Likewise, that's what molecules *feel like* in that configuration, and this is what they *feel like* in this configuration." We are again led back to a view of primary concepts: consciousness and content.

If consciousness doesn't need to combine in the way many have assumed it must for a panpsychic reality to be possible, then we don't face a combination problem at all. As we have seen, experiences of consciousness need not be continuous or maintained as individual selves or subjects. Nor do they necessarily need to be extinguished when the smaller constituents of matter combine to make more complex systems, like brains. The illusion of being a self, along with an experience of continuity over time through memory, may in fact be a very rare form of consciousness. Whatever the larger reality is, the particular experience we have is dictated by the structure and function of our brain, which may not offer us a helpful starting point for understanding the actual nature of consciousness. Is it possible that alongside the conscious experience of "me," there is a much dimmer experience of each individual neuron, or of different collections of neurons and cells in my body and beyond? Could the universe literally be teeming with consciousness flickering in and out, overlapping, combining, separating, flowing, in ways we can't quite imagine—depending on the laws of physics in a way we don't yet understand?

Perhaps the term "panpsychism," because of its history and associations, will continue to pose obstacles to progress in the field, and we need a new label for the work in which philosophers and scientists theorize about the possibility that consciousness is a fundamental feature of matter. Just as we have branches of physics, theoretical and experimental, we might need to come up with a new term for this branch of consciousness studies, distinguishing it from the work of neuroscientists who study the neural correlates of consciousness.[7]

Theories that entail panpsychism have been gaining respect in recent years, but they are still vulnerable to being nudged off the academic stage. In his article "Conscious Spoons, Really? Pushing Back against Panpsychism,"[8] Anil Seth expresses a common view among neuroscientists that consciousness science has "moved on" from grappling with Chalmers's "hard problem," and thus from such "fringe" solutions as panpsychism. He insists that "by building increasingly sophisticated bridges between mechanism and phenomenology, the apparent mystery of the hard problem may dissolve." However, the two lines of inquiry—attempting to understand which brain processes give rise to our human experience versus what consciousness is in the first place—can coincide, even if they may not necessarily inform each other. As is the case in physics, neuroscientists need not spend a moment studying theoretical ideas they're not interested in. But they need not get in the way of studying those ideas, either. The theoretical work in science is often a necessary starting point and just as vital to scientific progress as the experimental work that follows.

It's important to clarify a few points regarding the distinction I continue to draw between two categories of questions—those pertaining to how deep in the universe consciousness runs and those about the brain processes that give rise to our human experiences—along with the value I place on each of them. First, although I'm defending panpsychism as a legitimate category of theories about consciousness based on what we currently know, I am not closed to the possibility that we might discover, by some future scientific method, that consciousness does in fact exist only in brains. It's hard for me to see how we could ever arrive at this understanding with any certainty, but I don't rule it out. Nor am I discounting the possibility that consciousness is something we will never fully grasp. Rebecca Goldstein is likely right when she suggests that the mystery of consciousness is impervious to scientific methods:

> It is somewhat depressing to think of an absolute limit on our science: to know that there are things we can never know. . . . Mathematical physics has yielded knowledge of so many of the properties of matter. However, the fact that we material objects have experiences should convince us that it cannot, alas, yield knowledge of them all. Unless a new Galileo appears, who offers us a way of getting at properties of matter that need not be mathematically expressible, we will never make any scientific progress on the hard problem of consciousness.[9]

Furthermore, my focus is on the mystery presented by the hard problem of consciousness because I think it is underappreciated and needs our attention, especially as we face the wide array of artificial minds that will soon be among us. Understanding whether or not advanced AI is conscious is as important as any other moral question. You have an ethical obligation to call an ambulance if you find your neighbor critically injured, and you would suddenly have similar obligations toward artificially intelligent beings if you knew they were conscious. However, in discussions about less complex forms of consciousness that versions of panpsychism point to—such as those that might exist in a thermostat or an electron—concepts like happiness and suffering don't apply, and any intuition that the scope of our ethics extends to systems vastly different from ourselves seems premature. The questions in neuroscience that concern human suffering ("Is Amanda's experience successfully being put on hold under anesthesia?" for instance) are clearly the most urgent for scientists to address at this time.

It's also important to say, once again, that the two different lines of inquiry I have sketched out are not mutually exclusive, but they will probably always remain isolated from one another to a large degree. For example, we could discover that consciousness is ubiquitous, yet simultaneously know that one's *specific experience* ceases to exist under certain neurological conditions when that person is, in effect, unconscious—such as being in a coma or under anesthesia. Additionally, it seems probable that only complex minds are capable of great happiness and great suffering. In that case, even if a version of panpsychism were true, not all islands of consciousness would be equal, or equally important to understand. The fact remains that integrated and complex minds like ours are capable of immense suffering, and we should be motivated to help all beings avoid it whenever possible. And if we were limited to investigating only one path to solving the mystery of consciousness at a time (which, thankfully, we're not!), I would prioritize work like Anil Seth's and Giulio Tononi's. Nevertheless, the deeper mysteries are clearly worthy of continued scientific study and are currently in need of defense—to protect the values of curiosity and inquiry in our pursuit of knowledge. I agree with the conclusion that Murray Shanahan, a professor of cognitive robotics at Imperial College London, arrives at:

> To situate human consciousness within a larger space of possibilities strikes me as one of the most profound philosophical projects we can undertake. It is also a neglected one. With no giants upon whose shoulders to stand, the best we can do is cast a few flares into the darkness.[10]

It seems clear that the overall picture we currently have, along with the long list of questions lacking definitive answers, gives us good reason to keep thinking about consciousness in more creative ways—and specifically to continue entertaining the idea that consciousness goes deeper than our intuitions have led us to believe. Inquiry into the nature of consciousness, however, will move forward only if it is considered a mystery worthy of our curiosity.

8

CONSCIOUSNESS AND TIME

After ten minutes of practicing meditation together in silence, the second-grade students in my mindfulness class raised their hands to volunteer to share their experiences.[1] The first child to speak made a simple, yet profound, observation. "It's always the present moment, but there *is no present moment*. It keeps moving!" she exclaimed, excited by the sheer wonderment of her realization. It was delightful to watch her discover how the mystery of consciousness is related to the mystery of time: our awareness is experienced across time and cannot be separated from it.

Many neuroscientists have considered the possibility that the feeling we have of being in the present moment, with time continuously moving in one direction, is an illusion. In his book *Your Brain Is a Time Machine*, Dean Buonomano, a UCLA neuroscientist, explains that whether the flow of time is an illusion or a true insight into the nature of reality depends in part on which of these two opposing views in physics turns out to be correct:

1. Presentism: Time is in fact flowing and only the present moment is "real"; or
2. Eternalism: We live in a "block universe," where time is more like space—just because you are in one location (or moment) doesn't mean the others don't exist simultaneously.

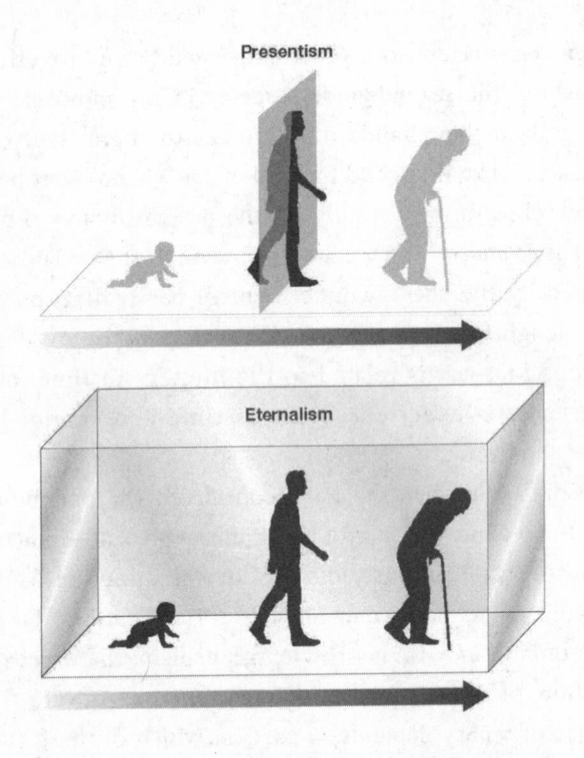

Figure 8.1: The two views of the nature of time.

Buonomano explains the difficulty of addressing the nature of time:

> These two views offer incompatible notions of the nature of time, but they both consider our feeling of the passage of time to represent a fundamental problem. Resolving this problem, however, will prove to be a formidable task, as our subjective sense of time sits at the center of a perfect storm of unsolved scientific mysteries: consciousness, free will, relativity, quantum mechanics, and the nature of time.[2]

In the perplexing world of quantum physics, John Wheeler's delayed-choice experiment—an experiment inspired by the results of the classic double-slit experiment—adds an even more mysterious layer to the question of how time relates to consciousness. In the double-slit experiment in quantum mechanics, when light is directed at a plate in which there are two parallel slits, the light acts like a wave—it passes through both slits, and results in an interference pattern on a screen placed behind the slits. This is true even if the light is emitted *one photon at a time* (Figure 8.2 [a]). This means that, somehow, an interference pattern is still created even though there is no actual interference we can detect between individual photons according to classical physics. It's as though each photon, wavelike, had passed through both slits simultaneously.

However, if a measurement is made at the slits to

determine which of them each individual photon passes through, the photons then act like particles, passing through one or the other of the slits and forming two parallel bands on the screen (as particles would be expected to do) and not the interference pattern (Figure 8.2 [b]).

This experiment tells us that light acts differently depending on whether or not it is being measured. Without a measurement, light acts like a wave; and when it is

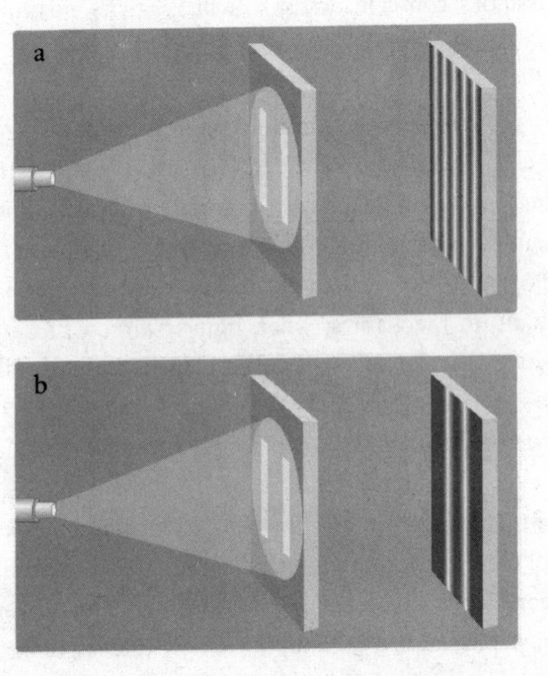

Figure 8.2: Double-slit experiments.

measured, it takes on the characteristics of individual particles. Some have made the claim that in order for light to behave like particles, not only does a measurement have to be made but that measurement has to be consciously observed. I'm not sure how anyone can definitively state that consciousness is implicated in the strangeness of the double-slit experiment, and here I follow the overwhelming consensus among scientists, including Wheeler: that photons exist in many possible states at once until interacting with *something*, but the something needn't be a conscious something. (This would change, of course, if we were to discover that consciousness is fundamental to matter, as consciousness would then be associated with all measurement, by definition.)

As if these results weren't strange enough, Wheeler introduced the element of time and made the prediction that even if we perform such a measurement *after* a photon has passed through one of the slits, we would still get the same effect, causing the photon to act like a particle *retroactively.*[3] In other words, he predicted that a measurement in the present would mysteriously influence the past. This is the delayed-choice experiment, and it was finally conducted in 2007, confirming Wheeler's prediction.[4]

Wheeler also proposed a related thought experiment in which he imagined measuring a single photon from the light emitted by a quasar billions of light-years away passing by a black hole on its way to a telescope on Earth. Just as in the double-slit experiment, the light would be split by the gravitational effect of the black hole, causing the phenomenon known as gravitational lensing—an optical illusion in which we see multiple images of a single source, such as a quasar. In an interview with the author Rob Reid, Donald Hoffman, a cognitive scientist at the University of California, Irvine, explains what would happen if we measured a single photon in Wheeler's cosmological thought experiment:

> You can now ask, for each photon that comes to me, whether it came from the left [or the right] side of the gravitational lens. [Let's say] I decide to measure which side it came from, and I find out that it went on the left side. That means I can say that for the last ten billion years, that photon has been on a path that started from the quasar and went around the left side of the gravitational lens. But if, instead, I had chosen not to make that measurement and just measure the interference pattern, it would not be true that for the last ten billion years that photon had gone [down a path] around the left side. So the choice I make today determines the ten-billion-year-history of that photon.[5]

In addition to the already incomprehensible facts that Wheeler's experiment reveals about light, if consciousness is in fact somehow intrinsic to matter, his experiment also suggests a very strange and counterintuitive relationship between consciousness and time.

Leaving the mind-twisting nature of quantum mechanics aside, let's return to the relative simplicity of our human experience in the present moment. Regardless of the actual nature of time, we know that our conscious experience doesn't represent the sequence of events in the world accurately. We have seen that through different processes the brain binds information that arrives at our sense receptors at different times and delivers it to us as a neat, present-moment package. But we can still wonder how *conscious experience itself* relates to time. Paying close attention to one's moment-to-moment experience through a concentration exercise like meditation—or simply contemplating the mystery of one's felt experience in general—leads to many interesting questions pertaining to time: How much time does a moment of consciousness take? Is consciousness continuous or does it somehow flicker in and out (and how would we know the difference)? What is the present moment; is it some sort of illusion? Is *time itself* an illusion?

Not only are all the questions surrounding consciousness important, especially as scientists and philosophers enter into the age of superintelligent machines, but they are fascinating to contemplate. In *The Tell-Tale Brain*, V. S. Ramachandran ponders the chances of science uncovering the mystery of consciousness: "Such advances could easily be as remote from our present-day grasp as molecular genetics was to those living in the Middle Ages. Unless there is a potential Einstein of neurology lurking around somewhere."[6]

From our current vantage point, it seems unlikely that we will ever arrive at a true understanding of consciousness. However, we may well be wrong about the absolute boundaries of knowledge. Humanity is young, and we've barely begun to understand our place in the cosmos. As we continue to look out from our planet and contemplate the nature of reality, we should remember that there is a mystery right here where we stand.

Acknowledgments

This book is the product of many years of research and lengthy conversations with experts in the field of consciousness studies. I am grateful to the scientists and philosophers who took time out of their demanding schedules to brainstorm (and debate at length!) with an amateur—it was an absolute joy to discuss consciousness with every one of them: Donald Hoffman, Anil Seth, Christof Koch, Rebecca Goldstein, Dean Buonomano, Philip Goff, Adam Frank, and Thomas Metzinger.

As this project evolved from an obsessive interest to a long article to a short book, many friends and colleagues played a critical role in its development. I am grateful for the generosity of all the scientists, philosophers, and artists alike who read and gave me notes on early drafts, sharing their curious minds and keen insights with me: Isabelle

Boemeke, Sean Carroll, David Chalmers, Antonio Damasio, Gavin de Becker, David Eagleman, Amy Eldon, Michael Gazzaniga, David Gelles, Joseph Goldstein, Daniel Goleman, Adam Grant, Susan Kaiser Greenland, Dan Harris, Nathalia Holt, Suzanne Hudson, Marco Iacoboni, David Janet, Amy Lenclos, Iain McGilchrist, Thomas Nagel, Rob Reid, Casey Rentz, Murray Shanahan, Jason Silva, Susan Smalley, Galen Strawson, Max Tegmark, Dalit Toledano, Giulio Tononi, Jon Turteltaub, Tim Urban, D. A. Wallach, Richelle Rich Waters, Diana Winston, and Kalika Yap. And a special thank-you to Gordon Gould for pulling me out of the panpsychism closet and getting me to finally put pen to paper.

Passionate thanks to Amy Rennert, the agent for my children's book, *I Wonder*, who supported this project from day one and whose instincts I always trust—this book wouldn't exist without her. To my agent, John Brockman, for taking a chance on a book about a crazy topic, and to Max Brockman for helping to convince him to take the leap. John and Katinka Brockman have been trusted friends for many years, and it would be impossible to list all the ways they have been an inspiration and a support to me. I am grateful not just for the opportunity to work with them but for the privilege of spending time with two people I so admire. To my editor and mentor, Sara Lippincott; I was greatly encouraged by her interest and her confidence, and the book is both more rigorous and more eloquent because of her invaluable input. I am also indebted to my editor at HarperCollins, Sarah Haugen, for her zeal and patience as we worked to get the execution of this complicated and controversial topic just right. She took great care to push me out of my comfort zone, and the book is much stronger as a result.

Our nanny, Rosmari, helped keep the world turning (and children from busting through the door of my home office) during the hours when I was researching and writing. My trust in her allowed me the freedom to continue the work I love—a luxury I am keenly aware that few women have had to this day, and for which I am deeply grateful.

Paul Witt, even as he was very ill, generously offered his expert feedback but sadly was not with us long enough to read the final manuscript. This book would have benefited tremendously from his talent and wisdom. We all miss him dearly, and his touch is surely missing in these pages.

My heartfelt thanks go to my sisters, Brianna and Jen, for always being willing readers, and who were available for feedback up to the eleventh hour (even offering last-minute edits via texts). I'm so fortunate to have their friendship and access to their natural editing skills. And to my mother for being my first and most devoted editor and for her unending support.

Last, and most of all, to Sam, Emma, and Violet, whose love is the most precious and cherished experience ever to arise in my consciousness.

Notes

Chapter 1: A Mystery Hiding in Plain Sight

1. Thomas Nagel, "What Is It Like to Be a Bat?," *The Philosophical Review* 83, no. 4 (1974): 435–50.

2. Rebecca Goldstein, "The Hard Problem of Consciousness and the Solitude of the Poet," *Tin House* 13, no. 3 (2012): 3.

3. The great mystery is usually phrased, "Why is there something rather than nothing?" But the more interesting question to me (and the question that is analogous to the hard problem) is: How *could something* come out of nothing? In other words, does it even make sense to ask the question? How do we even conceive of a process by which something is born out of nothing?

4. David Chalmers, "Facing Up to the Problem of Consciousness," *Journal of Consciousness Studies* 2, no. 3 (1995): 200–19. See also Galen Strawson, chapter 4, *Mental Reality* (Cambridge, MA: MIT Press, 1994): 93–96.

Chapter 2: Intuitions and Illusions

1. Ap Dijksterhuis and Loran F. Nordgren, "A Theory of Unconscious Thought," *Perspectives on Psychological Science* 1, no. 2 (June 2006): 95–109; Erik Dane, Kevin W. Rockmann, and Michael G. Pratt, "When Should I Trust My Gut?," *Organizational Behavior and Human Decision Processes* 119, no. 2 (November 2012): 187–94, https://doi.org/10.1016/j.obhdp.2012.07.009.

2. Liz Fields, "What Are the Odds of Surviving a Plane Crash?," ABC News, 12 March 2014, https://abcnews.go.com/International/odds-surviving-plane-crash/story?id=22886654.

3. Daniel Chamovitz, *What a Plant Knows: A Field Guide to the Senses* (New York: Farrar, Straus & Giroux, 2012), 68–69.

4. Gareth Cook, "Do Plants Think?," *Scientific American*, 5 June 2012, https://www.scientificamerican.com/article/do-plants-think-daniel-chamovitz/.

5. Suzanne Simard, "How Trees Talk to Each Other," TED talk, June 2016, www.ted.com/talks/suzanne_simard_how_trees_talk_to_each_other.

6. Nic Fleming, "Plants Talk to Each Other Using an Internet of Fungus," BBC News, 11 November 2014, http://www.bbc.com/earth/story/20141111-plants-have-a-hidden-internet; Paul Stamets, "6 Ways Mushrooms Can Save the World," TED talk, March 2008, https://www.ted.com/talks/paul_stamets_on_6_ways_mushrooms_can_save_the_world.

7. Lauren Goode, "How Google's Eerie Robot Phone Calls Hint at AI's Future," *Wired*, 8 May 2018, https://www.wired.com/story/google-duplex-phone-calls-ai-future; Bahar Gholipour, "New AI Tech Can Mimic Any Voice," *Scientific American*, 2 May 2017, https://www.scientificamerican.com/article/new-ai-tech-can-mimic-any-voice.

8. In other words, if consciousness comes at the end of a stream of information processing, does the fact that there is an experience make a difference to the brain processing that follows? Does consciousness affect the brain? See also Max Velmans, *How Could Conscious Experiences Affect Brains?* (Charlottesville, VA: Imprint Academic, 2002), 8–20.

9. Masao Migita, Etsuo Mizukami, and Yukio-Pegio Gunji, "Flexibility in Starfish Behavior by Multi-Layered Mechanism of Self-Organization," *Biosystems* 82, no. 2 (November 2005): 107–15, https://doi.org/10.1016/j.biosystems.2005.05.012.

Chapter 3: Is Consciousness Free?

1. David Eagleman, *The Brain: The Story of You* (New York: Pantheon, 2015), 53.

2. Electroencephalogram (EEG) is a noninvasive method of recording electrical activity in the brain through electrodes placed on the scalp.

3. See, for example, Chun Siong Soon, Anna Hanxi He, Stefan Bode, and John-Dylan Haynes, "Predicting Free Choices for Abstract Intentions," *Proceedings of the National Academy of Sciences* 110, no. 15 (April 2013) 6217–22; DOI: 10.1073/pnas.1212218110; Itzhak Fried, Roy Mukamel, and Gabriel Kreiman, "Internally Generated Preactivation of Single Neurons in Human Medial Frontal Cortex Predicts Volition," *Neuron* 69, no. 3 (February 2011): 548–62, https://doi.org/10.1016/j.neuron.2010.11.045;

Aaron Schurger, Myrto Mylopoulos, and David Rosenthal, "Neural Antecedents of Spontaneous Voluntary Movement: A New Perspective," *Trends in Cognitive Sciences* 20, no. 2 (February 2016): 77–79, https://doi.org/10.1016/j.tics.2015.11.003.

4. Quoted in Susan Blackmore, *Conversations on Consciousness* (New York: Oxford University Press, 2006), 252–53; see also Daniel Wegner and Thalia Wheatley, "Apparent Mental Causation: Sources of the Experience of Will," *American Psychologist* 54, no. 7 (July 1999): 480–92.

5. See, for instance, Daniel Wegner, *The Illusion of Conscious Will* (Cambridge, MA: MIT Press, 2003), 3–15.

6. For a fuller analysis of this issue, see, for instance, Sam Harris, *Free Will* (New York: Free Press, 2012).

Chapter 4: Along for the Ride

1. Kathleen McAuliffe, *This Is Your Brain on Parasites* (Boston: Houghton Mifflin Harcourt, 2016), 57–82.

2. McAuliffe, 79.

3. McAuliffe, 25–31.

4. Natalie Angier, "In Parasite Survival, Ploys to Get Help from a Host," *New York Times*, 26 June 2007, https://www.nytimes.com/2007/06/26/science/26angi.html.

5. Henry Fountain,. "Parasitic Butterflies Keep Options Open with Different Hosts," *New York Times*, 8 January 2008, https://www.nytimes.com/2008/01/08/science/08obmimi.html.

6. Mary Bates, "Meet 5 'Zombie' Parasites That Mind-Control Their Hosts," *National Geographic*, 2 November 2014, https://news.nationalgeographic.com/news/2014/10/141031-zombies-parasites-animals-science-halloween/.

7. Melinda Wenner, "Infected with Insanity," *Scientific American Mind*, May 2008, 40–47, https://www.scientificamerican.com/article/infected-with-insanity/.

8. "PANDAS—Questions and Answers," National Institute of Mental Health, NIH Publication No. OM 16-4309, September 2016, https://www.nimh.nih.gov/health/publications/pandas/pandas-qa-508_01272017_154202.pdf.

9. David Chalmers, *The Conscious Mind* (New York: Oxford University Press, 1996), 198–99.

Chapter 5: Who Are We?

1. Kathleen A. Garrison et al., "Meditation Leads to Reduced Default Mode Network Activity Beyond an Active Task," *Cognitive, Affective & Behavioral Neuroscience* 15, no. 3 (September 2015): 712, https://doi.org/10.3758/s13415-015-0358-3; Judson A. Brewer et al., "Meditation Experience Is Associated with Differences in Default Mode Network Activity and Connectivity," *Proceedings of the National Academy of Sciences* 108, no. 50 (13 December 2011): 20254–59, https://doi.org/10.1073/pnas.1112029108.

2. Robin Carhart-Harris et al., "Neural Correlates of the LSD Experience Revealed by Multimodal Imaging," *Proceedings of the National Academy of Sciences* 113, no. 17 (26 April 2016): 4853–58, https://doi.org/10.1073/pnas.1518377113.

3. Ian Sample, "Psychedelic Drugs Induce 'Heightened State of Consciousness,' Brain Scans Show," *Guardian*, 19 April 2017, https://www.theguardian.com/science/2017/apr/19/brain-scans-reveal-mind-opening-response-to-psychedelic-drug-trip-lsd-ketamine-psilocybin.

4. Michael Pollan, *How to Change Your Mind* (New York: Penguin Press, 2018), 304–5.

5. Erin Brodwin, "Why Psychedelics like Magic Mushrooms Kill the Ego and Fundamentally Transform the Brain," *Business Insider*, 17 January 2017, https://www.businessinsider.com/psychedelics-depression-anxiety-alcoholism-mental-illness-2017-1.

6. Pollan, *How to Change Your Mind*, 305.

7. Brodwin, "Why Psychedelics like Magic Mushrooms Kill the Ego and Fundamentally Transform the Brain."

8. Michael Harris, "How Conjoined Twins Are Making Scientists Question the Concept of Self," *The Walrus*, 6 November 2017, https://thewalrus.ca/how-conjoined-twins-are-making-scientists-question-the-concept-of-self/.

9. Andrew Olendzki, *Untangling Self* (Somerville, MA: Wisdom Publications, 2016), 2. Olendzki goes on to say on page 3: "There is no intrinsic identity in anything. There are only the labels we decide upon to refer to things: clouds, raindrops, puddles. All persons, places, and things are merely names that we give to certain patterns we call out from the incessant flux of interdependent natural events. Why are human beings any different from this? . . . Surely 'Joe' is just something that occurs when conditions come together in certain ways, and Joe no longer occurs when those conditions change enough. . . . Under some conditions Joe is living; when the conditions supporting Joe's life no longer occur, Joe will no longer be living. He

is not the sort of thing that can *go* somewhere else (to heaven or to another body, for example), except perhaps in the most abstract sense of the recycling of his constituent components. All this is as natural as a rainstorm in the summer."

10. The BrainPort invention belongs to a company called Wicab in Wisconsin.

11. Eagleman, *The Brain: The Story of You*,187.

12. David Eagleman, "Can We Create New Senses for Humans?," TED talk, March 2015, https://www.ted.com/talks/david_eagleman_can_we_create_new_senses_for_humans.

13. For more, see Olaf Blankee, "Out-of-Body Experience: Master of Illusion," *Nature* 480, no. 7376 (7 December 2011), https://www.nature.com/news/out-of-body-experience-master-of-illusion-1.9569; Ye Yuan and Anthony Steed, "Is the Rubber Hand Illusion Induced by Immersive Virtual Reality?," in *IEEE Virtual Reality 2010 Proceedings*, eds. Benjamin Lok, Gudrun Klinker, and Ryohei Nakatsu (Piscataway, NJ: Institute of Electrical and Electronics Engineers, 2010), 95–102.

14. Anil Seth, "Your Brain Hallucinates Your Conscious Reality," TED talk, April 2017, https://www.ted.com/talks/anil_seth_how_your_brain_hallucinates_your_conscious_reality.

15. See, for example, Iain McGilchrist, *The Master and His Emissary* (New Haven, CT: Yale University Press, 2009).

16. Christof Koch, *The Quest for Consciousness* (Englewood, CO: Roberts & Company, 2004), 287–94.

17. Koch, 292.

18. Michael Gazzaniga, "The Split Brain Revisited," *Scientific American*, July 1998, 54.

19. McGilchrist, *Master*, 220–21.

Chapter 6: Is Consciousness Everywhere?

1. The *Oxford English Dictionary* defines panpsychism as "the theory of belief that there is an element of consciousness in all matter." See also *Stanford Encyclopedia of Philosophy*, s.v. "panpsychism," revised 18 July 2017, https://plato.stanford.edu/entries/panpsychism/.

2. Philip Goff, "Panpsychism Is Crazy, but It's Also Most Probably True," *Aeon*, 1 March 2017, https://aeon.co/ideas/panpsychism-is-crazy-but-its-also-most-probably-true. Goff makes a strong case for a panpsychic view in this article and elsewhere, but many part company with him (myself

included) when he defends the hypothesis set out in his essay on "cosmo-psychism" ("Is the Universe a Conscious Mind?," *Aeon*, 8 February 2018, https://aeon.co/essays/cosmopsychism-explains-why-the-universe-is-fine-tuned-for-life) that "the Universe is conscious, and . . . the consciousness of humans and animals is derived not from the consciousness of fundamental particles, but from the consciousness of the Universe itself"—a universe that, Goff speculates, is an agent "aware of the consequences of its actions." The argument seems flawed to me, and Goff himself has had a change of heart, which he wrote about in a blog post on 24 April 2018: https://conscienceandconsciousness.com/2018/04/24/a-change-of-heart-on-fine-tuning/.

3. David Chalmers, "Strong and Weak Emergence," in *The Re-Emergence of Emergence: The Emergentist Hypothesis from Science to Religion*, eds. Philip Clayton and Paul Davies (New York: Oxford University Press, 2008).

4. David Skrbina, *Panpsychism in the West* (Cambridge, MA: MIT Press, 2017), 189–90. Galen Strawson also draws the conclusion that "there is no radical emergence." See "Physicalist panpsychism," in Susan Schneider and Max Velmans, eds., *The Blackwell Companion to Consciousness*, 2nd ed. (Hoboken, NJ: Wiley-Blackwell, 2017), pp. 384–85.

5. Skrbina, *Panpsychism in the West*, 194–95.

6. David Chalmers differentiates between "weak emergence" and "strong emergence." Describing weak emergence, Chalmers writes, "The 'emergent' properties are in fact deducible (perhaps with great difficulty) from the low-level properties, perhaps in conjunction with knowledge of initial conditions, so strong emergence [in the form of consciousness] is not at play here" (Chalmers, "Strong and Weak").

7. Galen Strawson, "The Consciousness Deniers," *NYR Daily* (blog), *New York Review of Books*, 13 March 2018, https://www.nybooks.com/daily/2018/03/13/the-consciousness-deniers/.

8. Blackmore, *Conversations on Consciousness*, 28.

9. Paradoxically, it seems to me that declaring consciousness to be an illusion is just one step away from asserting that everything is potentially conscious.

10. Galen Strawson, "Physicalist panpsychism," in Schneider and Velmans, eds., *The Blackwell Companion to Consciousness*, pp. 376–84.

11. V. S. Ramachandran, *The Tell-Tale Brain* (New York: W. W. Norton, 2011), 248.

12. Peter Hankins, "Francis Crick," *Conscious Entities* (blog), 9 August 2004, http://www.consciousentities.com/crick.htm. See also Francis Crick, *The Astonishing Hypothesis* (New York: Simon & Schuster, 1995), chap. 17.

13. "Zap and zip" is based on the work of Giulio Tononi's integrated information theory (IIT). See Giulio Tononi et al., "Integrated Information Theory: From Consciousness to Its Physical Substrate," *Nature Reviews Neuroscience* 17, no. 7 (July 2016): 450–61, https://www.nature.com/articles/nrn.2016.44.

14. Christof Koch, "How to Make a Consciousness Meter," *Scientific American*, November 2017, 28–30.

15. Steve Paulson, "The Spiritual, Reductionist Consciousness of Christof Koch," *Nautilus*, 6 April 2017, http://nautil.us/issue/47/consciousness/the-spiritual-reductionist-consciousness-of-christof-koch.

16. Ibid.

17. Chalmers, *Conscious Mind*, 294–95.

18. Even if we concede that it makes sense to view consciousness as an evolved function in aid of survival, the idea that a physical system could develop a property that is so un-material-like suggests to me that consciousness was there all along as a property to be called on by the physical system—which brings us back full circle to a version of panpsychism.

19. Adam Frank, "Minding Matter," *Aeon*, 13 March 2017, https://aeon.co/essays/materialism-alone-cannot-explain-the-riddle-of-consciousness.

20. Skrbina, *Panpsychism*, 9, 17.

21. Ibid., 235–36.

22. Gregg Rosenberg, "Rethinking Nature: A Hard Problem within the Hard Problem," in *Explaining Consciousness: The "Hard Problem,"* ed. Jonathan Shear (Cambridge, MA: MIT Press, 1997), 287–300.

Chapter 7: Beyond Panpsychism

1. Rebecca Goldstein, personal communication with author, 16 March 2018.

2. Galen Strawson, "Consciousness Isn't a Mystery. It's Matter," *New York Times*, 16 May 2016, https://www.nytimes.com/2016/05/16/opinion/consciousness-isnt-a-mystery-its-matter.html. See also Galen Strawson, "Consciousness Never Left," in K. Almqvist and A. Haag, eds., *The Return of Consciousness: A New Science on Old Questions* (Stockholm: Ax:son Johnson Foundation, 2017): 87–103. Strawson and others also prefer to state the

mystery in terms of *why* consciousness exists, as opposed to *what* it is. I have gone back and forth about how to phrase the mystery myself. The problem I have with posing the question in terms of *why* is its religious undertone. It also evades the hard problem by inviting the ready response, "Well, of course the reason we're conscious is because our neurons are doing this special thing that causes us to be conscious." When I put it in terms of *what*, however, I mean, "*What* causes consciousness? *What* is the overall big-picture explanation?" The *what* question also more readily opens people's minds to all the follow-up questions: Is consciousness intrinsic to matter? Where does it come from? What exactly is it from top to bottom?

3. Skrbina, *Panpsychism*, 260.

4. *Stanford Encyclopedia of Philosophy*, s.v. "panpsychism," https://plato.stanford.edu/entries/panpsychism/#OtheArguForPanp.

5. David Chalmers, "The Combination Problem for Panpsychism," in *Panpsychism: Contemporary Perspectives*, eds. Godehard Bruntrup and Ludwig Jaskolla (New York: Oxford University Press, 2003).

6. See also William Hirstein, *Mindmelding: Consciousness, Neuroscience, and the Mind's Privacy* (New York: Oxford University Press, 2012).

7. One example that would fall into this category is a new theory Donald Hoffman is developing called "conscious realism." His theory rests on the idea that while evolution selects for fitness in organisms, it does not select for perceptions that present us with the truth about the fundamental nature of reality. According to Hoffman's work, in order for evolution by natural selection to effectively select for fitness, it must actually select *against* the perception of reality as it is. Therefore, everything we perceive, including space and time, is an incorrect view of the deeper fundamental reality in which we exist. Hoffman therefore argues that the fundamental components of reality cannot be described in terms of physical matter in space-time but instead must be a form of consciousness with interacting systems he has termed "conscious agents." Whether or not Hoffman's current theory turns out to be correct, his work is scientifically rigorous and offers a promising line of research that may at least help us grab a foothold where we would otherwise seem to have no hope of gaining any ground—and, in the meantime, he is, hopefully, pushing against the limits of our intuitions and expanding the possibilities of how we are willing to think about the universe. See Donald Hoffman, *The Case Against Reality: Why Evolution Hid the Truth from Our Eyes* (New York: W. W. Norton & Company, 2019).

8. Anil Seth, "Conscious Spoons, Really? Pushing Back against Panpsychism," *NeuroBanter* (blog), 1 February 2018, https://neurobanter.com/2018/02/01/conscious-spoons-really-pushing-back-against-panpsychism/.

9. Rebecca Goldstein, "Reduction, Realism, and the Mind" (PhD dissertation, Princeton University, 1977), with an addition from a personal communication with the author, 16 March 2018.

10. Murray Shanahan, "Conscious Exotica: From Algorithms to Aliens, Could Humans Ever Understand Minds That Are Radically Unlike Our Own?," *Aeon*, 19 October 2016, https://aeon.co/essays/beyond-humans-what-other-kinds-of-minds-might-be-out-there.

Chapter 8: Consciousness and Time

1. I was trained to teach mindfulness meditation to children by Susan Kaiser Greenland, and I have been volunteering for Greenland's Inner Kids foundation since 2005. See https://www.susankaisergreenland.com.

2. Dean Buonomano, *Your Brain Is a Time Machine* (New York: W. W. Norton, 2017), 216.

3. John A. Wheeler, "Law Without Law," in *Quantum Theory and Measurement*, eds. John A. Wheeler and Wojciech H. Zurek (Princeton, NJ: Princeton University Press, 1984), 182–213.

4. Vincent Jacques et al., "Experimental Realization of Wheeler's Delayed-Choice Gedanken Experiment," *Science* 315, no. 5814: 966–68, 16 February 2007, https://doi.org/10.1126/science.1136303.

5. Rob Reid and Donald Hoffman, "The Case against Reality," *After On* (podcast), episode 26, 30 April 2018; see also John A. Wheeler, "Law Without Law," 190.

6. Ramachandran, *Tell-Tale Brain*, 249.

Index

Page references in *italics* refer to illustrations.

About the Author

Annaka Harris is an author, an editor, and a consultant for science writers. She is the author of the children's book *I Wonder* and a collaborator on Susan Kaiser Greenland's Mindful Games Activity Cards, and her work has appeared in the *New York Times*. She lives with her husband, the neuroscientist, author, and podcaster Sam Harris, and their two children.

WITHDRAWN FROM STOCK
DUBLIN CITY PUBLIC LIBRARIES